FOODPLANTS OF WORLD SATURNIIDAE

FRONTISPIECE

1. ***Citheronia beledonon*** Dyar, MEXICO: Cuernavaca, on *Liquidambar styraciflua* (HAMAMELIDACEAE) (photo, Simon, L.).
2. ***Eochroa trimeni*** Felder, SOUTH AFRICA: Namaquland, on *Melianthus dregianus* (MELIANTHACEAE) (photo, Wolfe, K. L.).
3. ***Actias truncatipennis*** (Sonthonnax), MEXICO: Veracruz near Las Minas, on *Liquidambar* (HAMAMELIDACEAE) (photo, Wolfe, K. L.).
4. ***Polythysana apollina*** Felder, CHILE: Linares, on *Cryptocarya rubra* (LAURACEAE) (photo, Wolfe, K. L.).
5. ***Arsenura*** Duncan sp., COSTA RICA: Río Negro, on *Curatella americana* (DILLENIACEAE) (photo, Wolfe, K. L.) (foodplant record received after publication went to press).

FOODPLANTS OF WORLD SATURNIIDAE

Stephen E. Stone
18102 East Oxford Drive
Aurora, Colorado 80013 USA

The Lepidopterists' Society
Memoir Number 4, 1991
William E. Miller, Editor

For My Parents

Library of Congress Catalog Number: 91-62071
ISBN: 0-930282-05-1.
Printed in the United States of America.

Cover illustration: *Citheronia regalis* larva on *Syringa vulgaris* from an original transparent water color by John Cody.
Cover design by Stephen E. Stone.

CONTENTS

FOREWORD

Contrary to a widely held opinion, rearing Lepidoptera is not only an avocation of amateur entomologists and the most convenient process to get absolutely perfect specimens for pinning. Knowledge of the early stages is of the highest importance for the taxonomist as a source of information as useful as the study of adult morphology, from the specific to the suprageneric level. Scientific observations on living material reveal significant characters that can be crucial for taxonomic decisions; they are also of high significance in phylogeny.

As chances to make these observations in the wild, especially regarding exotic species, are unfortunately few, the only way to observe living insects and to know more about their biology is to rear them. Rightly or wrongly, silkmoths are reported to be relatively easy to rear and most of them will reproduce in captivity. This, in addition to their spectacular appearance, is among the reasons for their popularity, and it explains why they are probably the most commonly reared Lepidoptera.

The rearer has to face two main problems, each of about equal importance. The first consists in recreating for the insect, at the different stages, ecological conditions as much as possible resembling those in nature; the second is to supply the larvae as soon as hatched with plants in accordance with their feeding requirements.

Searching for and selecting acceptable foodplants becomes a rather ticklish matter as soon as one attempts to rear nonindigenous species or exotic ones, which are not among the most commonly available from dealers. When larvae hatch, it is necessary to find out how to feed them in the shortest time, at any rate within the first 24 hours. At this point it is too late to rush to libraries to consult a wide range of literature, and there is no other method than to supply the young larvae with a bouquet of foodplant choices and hope for the best.

Such hypothetic researches and hazardous experiments will be in most cases avoided thanks to a comprehensive work as easy to use as the one presented to us here by Stephen E. Stone under the auspices of The Lepidopterists' Society. Over 500 species and subspecies, more than one-third of the known Saturniidae, are cited with their foodplants; thus I have learned that more than 250 species can be reared on plants to which I easily have access in my region.

This world-level compilation necessarily required quite an amount of meticulous bibliographical research; the impressive number of references is in this respect quite significant. However, this work is not only a compilation; many of the previously unpublished records are directly attributable to the author, himself an experienced breeder, and to his own observations of the North American Saturniidae in their natural habitats.

Of course, as explained by the author himself in the Introduction, it could not be expected to have at the first go an absolutely complete account of all reared species and their foodplants, mostly owing to sparseness of the publications. Moreover, many breeders are unaware of the interest of their observations and do not take care to record them; so many data have never been and never will be published. But, such as it is, the catalog is extremely substantial, and it seems improbable that any really important publication was overlooked.

A critical examination would probably reveal a few inexactitudes in the family-group taxonomic treatment, but this is quite excusable because the nomenclature is still wavering in many genera of Saturniidae; the present work would have remained unpublished for many years if delayed for the completion of the desirable revisions.

If reservations were to be expressed, they would bear on the absence of distinction between hosts in captivity and foodplants in the wild, especially when these are known with certainty, as from recent publications by Daniel Janzen on the Saturniidae of Costa Rica to which the author broadly refers. These two categories of hosts are quite different from an ecological and evolutionary point of view; that somebody has once successfully reared one species on an occasional host in the laboratory is of little significance if compared with a foodplant record in nature. Moreover, rearers very often forget to specify if captive foodplants have produced adults, and to report the percentage of mortality.

These remarks do not diminish the value of a publication that will certainly remain the saturniid rearer's reference manual for many years. This world-level account of foodplants of saturniid larvae will be useful to biologists and to ecologists as well as to amateur entomologists, and it will definitely stimulate scientific research in a still underexplored field of entomology.

I take great pleasure in warmly congratulating Stephen E. Stone, and wish to thank The Lepidopterists' Society for the initiative it has taken in publishing this fine work.

Claude Lemaire
Gordes, France

PREFACE

The embryo for this "catalog of larval cuisine" germinated some time ago and consisted of a few notes jotted down on odd scraps of paper. I have always loved nature and all things natural, and I was fortunate as a child inasmuch as my parents recognized this predilection and did everything possible to nurture what eventually became a career in the natural sciences. With a 1978 move to Denver, Colorado, and an immediate discovery of a huge urban population of *Hyalophora cecropia*, my general insect collection experienced a rapid evolution focussing exclusively on the Saturniidae. I wanted to learn as much as possible about this exquisite group of moths, but was frustrated because I did not even know where to start. Unquestionably, the winter of 1979 was a most fortuitous period for me — I met, via correspondence, Richard S. Peigler, then a Ph.D. candidate at Texas A&M University. Without delay, Ric informed me of an organization whose membership was concerned with study of Lepidoptera, The Lepidopterists' Society, an organization I promptly joined. Within a few months of my initial contact with Ric, I was introduced to Michael J. Smith, an Air Force major stationed in Tucson, Arizona. Both of these gentlemen have considerable experience collecting, rearing, and studying saturniid moths. With meticulous patience and painstaking detail both Ric and Mike explained there was more, much more, to conducting research on Saturniidae than my then "stamp collector" approach — a bunch of pretty silkmoths neatly arranged in a display case — and one of the best ways to acquire specimens was to rear them. Rear saturniid moths, what a novel idea; actually plant trees and shrubs in my back yard for the sole purpose of rearing moth larvae — build an insectary! My fires were fueled and a corner of the yard selected.

Next came the question of what to plant. I hoped to rear saturniids from various parts of the world, which translated directly into a variety of foodplant species. In my quest to define which foodplants to cultivate, I consulted Crotch (1956), Villiard (1969), Ferguson (1971-72), and other references focussing on moth biology, rearing, and foodplants. Based on this research, it became apparent that larval Saturniidae feed on a considerable assortment of primary foodplants, and many saturniids will accept an even broader spectrum of alternate foodplants. This was valuable information because Denver, Colorado, is in the high plains of the Front Range physiographic province just east of the southern Rocky Mountains. Environmentally, Denver is a high plains desert with winter temperatures not uncommonly recorded between -30°C. and -35°C. Snow and freezing temperatures have been recorded as late as early June and as early as the second week of September. These winter temperatures precluded flora that could not tolerate cold climates, such as tropical and hot desert species. In order to rear saturniids from these habitats, I either had to build a very large greenhouse in which to raise primary foodplants or find alternate foodplants that would survive in a high plains desert environment. Alternate foodplants are much easier and less expensive to purchase and maintain than a large greenhouse. This mandated that the insectary be designed with alternate foodplant species as a priority. Therefore, I selected, for the most part, species

I consider rather generic foodplants upon which I could hopefully rear saturniids from the world over. These plants include various species of *Ceanothus* and *Rhamnus* (RHAMNACEAE), *Crataegus, Prunus*, and *Pyrus* (ROSACEAE), *Fraxinus, Ligustrum,* and *Syringa* (OLEACEAE), *Gleditsia* (FABACEAE: CAESALPINIOIDEAE), *Pinus* (PINACEAE), *Populus* and *Salix* (SALICACEAE), *Rhus* (ANACARDIACEAE), *Tilia* (TILIACEAE), *Viburnum* (CAPRIFOLIACEAE), and, of course, the ever present *Quercus* (FAGACEAE). Occasionally I have an opportunity to rear saturniid species that will feed only on peculiar foodplants that will not survive outdoors, and for those special occurrences I baby the larvae along on potted plants maintained in my study.

I collected and reared local saturniids and exchanged livestock with other members of The Lepidopterists' Society. Rearing saturniids is a fascinating venture, and I began keeping personal notes on the species I reared — what foodplants various species would accept. I also started a serious search of the literature seeking more records of saturniid foodplants. The notes were pinned to a cork bulletin board in my study. In a short time the bulletin board was obscured with notes and pins so the notes were sorted, organized, and incorporated into a notebook. Writing notes in a permanent log is a fine way to keep information, but with the addition of copious new records it very quickly became difficult to find specific records of who eats what with any degree of alacrity. Enter the computer age. All my records were entered into a "Bugfood" file where they could be alphabetized, indexed, and quickly found through the search feature. After a few years the file grew quite large and many fellow rearers were seeking my advice on what foodplants to offer various saturniids. It occurred to me it was time to publish this compilation.

Stephen E. Stone
Aurora, Colorado

ACKNOWLEDGMENTS

A comprehensive catalog such as this would not be possible without the collective efforts of many individuals. Thanks must be expressed to those from whom I received personal communication and freely shared rearing information and assisted with literature searches, especially B. C. Kondratieff (Colorado State University, Fort Collins, Colorado), B. Lym (California Academy of Sciences, San Francisco, California), S. Naumann, P. A. Opler (U. S. Fish and Wildlife Service, Fort Collins, Colorado), R. S. Peigler (Denver Museum of Natural History, Colorado), D. C. Sblendorio, M. J. Smith (Research Associate, Nevada State Museum and Historical Society, Las Vegas, Nevada), J. P. Tuttle, B. D. Williams, and K. L. Wolfe (Field Associate, San Diego Natural History Museum, San Diego, California). All these individuals very kindly devoted considerable time and effort to this treatment. For reviewing the manuscript and offering creative suggestions on content, format, and style, gratitude is expressed to J. N. Borland, B. A. Drummond (Natural Perspectives, Woodland Park, Colorado), C. L. Fischer and J. F. Haman (National Park Service [NPS], Denver, Colorado), C. Lemaire (Correspondent de Muséum national d'Histoire Naturelle, Paris), J. E. Rawlins (Carnegie Museum of Natural History, Philadelphia, Pennsylvania), M. J. Smith, P. M. Tuskes, and an anonymous reviewer. Appreciation is extended to G. F. Russell (U.S. National Herbarium, Washington, D. C.), J. N. Borland, and S. G. Gignac (Denver Botanic Gardens, Colorado), for assistance in locating botanical references. I am particularly indebted to J. E. Rawlins for the many hours he contributed to ensuring the botanical nomenclature is the most current and accurate as possible. I am extremely grateful to L. D. Simon and K. L. Wolfe for providing the excellent photographs illustrated in the Frontispiece. I am deeply indebted to J. Cody for the superb cover illustration. The expertise and endurance of G. A. Jarvis (NPS), a veritable computer wizard, must be acknowledged for, in a very real sense, this catalog would not be possible without his good help. Sincere appreciation is also extended to E. A. Janes and M. L. Miller (NPS) for their continuing support. Finally, I thank my wife, K. J. Stone, for her enduring patience and sensitivity during the long hours I spent on the computer preparing this catalog.

INTRODUCTION

This effort is an attempt to compile a compendium of known foodplants of world Saturniidae, including laboratory-produced intergeneric and intrageneric hybrids. The family Saturniidae is an immensely popular group of Lepidoptera intensively studied by workers throughout the world. Both seasoned veterans and fresh beginners find innumerable rewards from rearing these fascinating animals. Many saturniid species are well known, and living material is easily accessible through personal collecting, exchanges with others, or from a variety of commercial sources. The biology of the more popular species has been chronicled through decades of intensive study and research, and many of their foodplants are well documented in the literature. Living material of other species is less frequently available, especially those species found in the more remote or inaccessible areas of the world. The documented biology of these latter species is correspondingly limited.

Foodplant sources in this catalog include published records and personal experiences of the author and others. Only foodplants resulting in at least one adult moth are provided. Published citations and personal observations of larvae *rarely* feeding on a specific foodplant are also provided. Such foodplant citations are preceded with a "?".

Saturniid foodplant records have been published in many languages in an assortment of journals, monographs, and other publications, some of which are obscure, difficult to locate, and more often than not discuss a single species or small group of species. Based largely on Crotch (1956), Gardiner (1982) compiled a list of world saturniid foodplants. That list contained much information, but many references published before that date were overlooked (or not cited) and considerably more saturniid foodplant information has accumulated since that time. Together with unpublished records, the present catalog integrates information from recent publications as well as from older, relatively obscure papers.

It is not the purpose of this catalog to produce a bibliographic recital of all authors citing the same foodplant for individual saturniid species. Such treatment would be redundant, consume enormous space, unnecessarily burden the "References Consulted" section, and detract from the principal theme — to provide a useful, readily accessible foodplant guide for rearing saturniid larvae. Further, no attempt has been made to substantiate or refute the validity of published foodplant records. Such documentation would require rearing each saturniid species on every recorded foodplant. Obviously, rearing on such a grand scale is not possible. However, it is important to remember that not all larvae from all populations will always accept each foodplant listed in this catalog. As an example, in nature *Hyalophora columbia columbia* larvae are, for the most part, found only on *Larix laricina*, a deciduous conifer. However, on occasion, in the laboratory, I have been able to rear *some* individuals of *some* populations of *H. columbia columbia* larvae to adults on *Prunus, Salix*, and *Pseudotsuga*.

This catalog is divided into two main parts. The first is arranged alphabetically by saturniid genus and species, each species followed by known larval foodplants. Added to this is a list of intergeneric and intrageneric saturniid hybrids and the foodplants on which they were reared. The arrangement of foodplants listed after each saturniid species is not random as appearance would suggest. While conducting research for this catalog, I noted that each author had a distinct way of listing saturniid foodplants. As an example, Tietz (1972) and Arora & Gupta (1979) alphabetized foodplant species, Janzen (1984a) mostly lists foodplants alphabetically by family, Ferguson (1971-72) effectively weaves the list of foodplants into the general discussion of moth biology, as do Cryan & Dirig (1977). Occasionally, however, Ferguson (1971-72) fluctuates between apparent randomization and listing foodplants in approximate order of larval acceptance as in his discussion of *Antheraea polyphemus polyphemus*. Crotch (1956) and Gardiner (1982) cite foodplants in what would appear to be a truly random order. In this present catalog, foodplant citations following each saturniid species closely follow the order of each cited author. The second main part is wholly alphabetical in arrangement and is arranged according to genus and species of foodplants followed by the saturniids that are known to feed on each species.

No attempt has been made to distinguish between foodplants found in nature and alternates used in the laboratory. Although it would be useful, many foodplants cited by authors as being used in nature are, in fact, human-introduced "exotic" species that native saturniids have exploited and would not otherwise use. While many references consulted for this catalog do distinguish between foodplants used in nature and those used in captivity, other publications do not, and it is beyond the scope of this effort to speculate between laboratory or naturally occurring saturniid foodplants.

During research for this catalog it became apparent that some authors cite foodplants using only vernacular names. Many such names are local or regional in origin and are not associated in print with scientific names. Because of this, Appendix I has been furnished to provide the more frequently used vernacular names of some saturniid foodplants. It is not always possible to do this because many plants may not have a vernacular name or if a vernacular name exists, its use is local and not always in print. To further confound the vernacular name problem, often a single species or genus will have different vernacular names in different parts of the world, or a vernacular name will encompass a whole genus, a number of genera, or an entire family — hence the value of the scientific name to identify a specific organism. Vernacular names of foodplants have been obtained from a variety of publications, which are listed in the References Consulted section.

Taxonomy is a dynamic science and research brings change. Regarding saturniid subspecies, a particular subspecies may not have always been cited by previous authors or those from whom I received personal communications, or new subspecies may have been described since a foodplant record was cited for the species in question. As an example, Packard (1914), Villiard (1969), and Ferguson (1971-72) discuss *Rothschildia orizaba* and provide foodplant records. Lemaire (1978) established the *R. orizaba* species-group which includes *R. orizaba orizaba, R. orizaba uruapana, R. orizaba triloba, R. orizaba equatorialis, R. orizaba cauca, R. orizaba peruviana*, and *R. orizaba meridana*. *Rothschildia orizaba* records prior to Lemaire (1978) are collectively presented as "*Rothschildia orizaba* (Westwood) ssp." I have used this unified approach for all saturniids where the historic subspecific identity is questionable. Additionally, in some instances, fairly recent taxonomic revision may incumber readily locating some saturniid species previously well known by a historically used if not incorrect name; such revisions include subspecies, species, and, in a few instances, an entire genus. Since previously published foodplant records cite these long-used and familiar names and this present work follows the current taxonomy, it seems worthwhile to provide direction for revised names within the text. This is accomplished by listing the species under its familiar name followed by the most current name and the citation from which the current name and synonymy may be traced. An example is "*Automeris aurantiaca* Weymer - See *Automeris naranja* Schaus ssp. (Lemaire, 1971-74: 300)."

Instructions or advice on how to acquire, collect, or rear saturniid moths is not provided. This information is well chronicled in numerous other references such as Galtsoff *et al.* (1937), Villiard (1969), Riotte & Peigler (1981), Gardiner (1982) (which is based heavily on Crotch, 1956), Peigler (1989), and would needlessly take up space in this catalog.

Omissions of published records will undeniably be found. However, as with any bibliographic investigation, a compilation of every saturniid foodplant record ever published would involve a search comprising several lifetimes. A line of demarkation has to be drawn somewhere, and this catalog includes all references (both in the printed record and personal communications) that I could find through a search of several years. Obviously, a list such as this will never be complete; it is assumed a continuing search through existing and forthcoming literature as well as input from workers rearing saturniids will constantly generate new foodplant records.

Nomenclature of New World Saturniidae follows the work of Lemaire (1971-74, 1978, 1980, & 1988). Nomenclature of Old World Saturniidae generally follows Bouvier (1936) with some revisions by Lemaire (*pers. comm.*).

Finally, the United States Department of Agriculture (USDA) regulates the importation of living organisms, including saturniid material, from outside the United States. Such importation requires the recipient to hold a valid import permit. Before receiving living material from outside the United States it is advised to contact your local USDA office for the proper permits.

PART 1
SATURNIIDS AND THEIR FOODPLANTS

ACTIAS Leach

artemis Bremer - ***Juglans regia, Quercus, Prunus*** (Crotch, 1956); ***Carya*** (Gardiner, 1982); ***Alnus firma, Betula platyphylla, Fagus crenata, Rosa*** (Inoue *et al.*, 1982); ***Castanea mollissima, Salix, Prunus pseudo-cerasus*** (Zhu & Wang, 1983); ***Pyrus malus, P. communis, Prunus armeniaca, Ilex verticillata, Acer, Punica*** (Albaugh, 1986); ***Prunus virginiana*** (Stone, S.E., pers. obs.).

dubernardi (Oberthür) - ***Pinus massoniana*** (Mell, 1950).

luna (Linnaeus) ssp. - ***Juglans cinerea, Carya*** (Morris, 1862); ***Liquidambar styraciflua, Diospyros virginiana*** (Holland, 1903); ***Fagus*** (Forbes, 1923); ***Juglans nigra, Quercus, Platanus lindeniana, Salix caprea, Prunus, Castanea sativa*** (Crotch, 1956); ***Betula papyrifera, Carpinus, Quercus rubra, Q. alba, Ulmus, Populus tremuloides, Salix, Acer saccharum, A. rubrum, Corylus, Alnus, Tilia*** (McGugan *et al.*, 1958); ***Carya cordiformis*** (Wright, 1967); "various JUGLANDACEAE, including" ***Carya illinoensis*** (Ferguson, 1971-72); ***Castanea dentata, Humulus lupulus, Nyssa sylvatica, Ostrya virginiana, Prunus domestica*** (Tietz, 1972); ***Juglans regia, Populus, Acer*** (Gardiner, 1982); ***Betula pendula*** (Naumann, S., pers. comm.); ***Carya glabra, Rhus glabra*** (Tuttle, J.P., pers. comm.); ***Carya ovata*** (Williams, B.D., pers. comm.).

maenas Doubleday - ***Turpina pomifera*** (Arora & Gupta, 1979); ***Eucalyptus gunnii, Castanea*** (Gardiner, 1982); ***Averrhoa bilimbi, Adinandra dumosa, Liquidambar styraciflua, Rhus typhina, R. copallina, R. glabra, R. radicans, Quercus, Turpina sphaerocarpa, Schima wallichii*** (Nässig & Peigler, 1984); ***Canarium*** (Holloway, 1987).

selene (Hübner) ssp. - ***Salix fragilis*** (Packard, 1914); ***Prunus, Crataegus*** (Crotch, 1956); ***Terminalia crenulata*** (Deodikar *et al.*, 1969); ***Azadirachta indica, Cedrela paniculata, Corylus colurna, Lannea coromandelica, Lawsonia alba, Moringa oleifera, Prunus amygdalus, P. cerasus, P. padus, Rhamnus frangula, Salix babylonica, Zanthoxylum acanthopodium, Z. alatum*** (Arora & Gupta, 1979); ***Cinnamomum camphora, Sapium sebiferum*** (Allen, 1981); ***Mangifera, Juglans regia, Pyrus malus, Salix, Hibiscus, Coriaria nepalensis, Andromeda ovalifolia, Rhododendron, Quercus*** (Gardiner, 1982); ***Andromeda glaucophylla, Liquidambar styraciflua, Syringa vulgaris, Ligustrum, Pyrus communis*** (Albaugh, 1986); ***Terminalia*** (Holloway, 1987); ***Castanea*** (Heppner *et al.*, 1988); ***Alnus glutinosa*** (Naumann, S., pers. comm.); ***Rhus laurina*** (Wolfe, K.L., pers. comm.).

selene gnoma Butler - ***Alnus japonica*** (Inoue *et al.*, 1982).

selene ningpoana Felder - ***Quercus ilex*** (Crotch, 1956); ***Pterocarya stenoptera, Salix, Castanea mollissima, Sapium sebiferum, Prunus armeniaca, Malus pumila, M. asiatica, Pyrus communis, Alnus cremastogyne, Hibiscus syriacus*** (Zhu & Wang, 1983); ***Alnus formosana*** (Heppner *et al.*, 1988); ***Liquidambar formosana, Prunus pseudo-cerasus, Cinnamomum camphora, Lagerstroemia subcostata*** (Wang, 1988).

sinensis heterogyna Mell - *Liquidambar formosana, L. styraciflua* (Nässig & Peigler, 1984).

truncatipennis (Sonthonnax) - *Liquidambar* (Beutelspacher, 1978); JUGLANDACEAE (Gardiner, 1982); *Juglans* (Wolfe, K.L., pers. comm.).

ADELOCEPHALA Travassos

brevis Walker - See ***Oiticella brevis*** (Walker) (Lemaire, 1988: 431).

crocata Boisduval - See ***Cicia crocata*** (Boisduval) (Lemaire, 1988: 189).

wardii (Boisduval) - See ***Adelowalkeria flavosignata*** (Walker) (Lemaire, 1988: 377).

ADELONEIVAIA Travassos

boisduvali (Doumergue) - *Acacia dealbata* (Herbin, D., pers. comm.).

fallax (Boisduval) - *Inga* (Lemaire, 1988).

isara (Dognin) - *Lysiloma divaricata, L. auritum* (Janzen, 1984a).

jason (Boisduval) ssp. - *Quercus agrifolia, Q. dumosa, Q. kelloggii* (Wolfe, K.L., pers. comm.).

subangulata (Herrich-Schaeffer) ssp. - "A spiny *Mimosa*" (Packard, 1905); *Salix capraea* (Lemaire, 1988).

ADELOWALKERIA Travassos

flavosignata (Walker) - "A spiny *Mimosa*" (Packard, 1905); *Machaerium brasiliensis* (Lemaire, 1988).

ADETOMERIS Michener

erythrops (Blanchard) - *Fagus pumilio, Maytenus boaria, Rubus ulmifolius* (Gardiner, 1982); *Rhus laurina, Prunus armeniaca, Fraxinus, Rhamnus* (Wolfe, K.L., pers. comm.).

AGAPEMA Neumoegen & Dyar

galbina Clemens ssp. - *Ribes* (Crotch, 1956); *Condalia hookeri, Ligustrum, Prunus* (Collins & Weast, 1961); *Celtis* (Gardiner, 1982).

galbina anona (Ottolengui) - *Sarcobatus maximiliani* (Packard, 1914); *Lycium* (Weast, 1962); *Condalia spathulata* (Ferguson, 1971-72).

galbina galbina (Clemens) - *Condalia ericoides* (Tuttle, J.P., pers. comm.).

homogena Dyar - *Rhamnus californicus, Salix* (Tuskes & Smith, 1984); *Ribes cereum* (Evans, H.E., pers. comm.); *Rhamnus frangula* (Naumann, S., pers. comm.); *Ribes alpinum* (Stone, S.E., pers. obs.).

solita Ferguson - *Condalia obovata* (Ferguson, 1971-72).

AGLIA Ochsenheimer

tau (Linnaeus) ssp. - *Fagus, Betula, Platanus, Acer pseudoplatanus* (Crotch, 1956); *Fagus sylvatica, Quercus robur, Carpinus betulus, Betula alba, Corylus avellana, Pyrus communis, Salix capraea* (Rougeot, 1971); "most deciduous trees, esp."

Pyrus malus, Tilia (Gardiner, 1982); ***Alnus japonica, Castanea crenata, Fagus crenata, Quercus dentata, Q. serrata*** (Inoue *et al.*, 1982); ***Corylus heterophylla, Alnus cremastogyne, Betula platyphylla, Fagus longipetiolata*** (Zhu & Wang, 1983); ***Juglans regia, Ilex verticillata*** (Albaugh, 1986); ***Betula pendula*** (Naumann, S., pers. comm.).

ANISOTA Hübner

"Larvae of all species eat *Quercus* and some will accept *Castanea*" (Riotte & Peigler, 1980).

assimilis (Druce) - *Quercus grisea* (Riotte & Peigler, 1980); *Quercus albocincta* (Smith, M.J., pers. comm.).

consularis Dyar - *Quercus virginiana* (Ferguson, 1971-72); *Quercus nigra, Q. falcata, Q. laevis, Q. myrtifolia* (Riotte & Peigler, 1980); *Quercus palustris* (Tuttle, J.P., pers. comm.).

dissimilis (Boisduval) - *Quercus* (Schaus, 1884).

finlaysoni Riotte - *Castanea vulgaris, Quercus muehlenbergii, Q. alba, Q. rubra, Q. prinus* [1] (Riotte & Peigler, 1980); *Quercus palustris* (Tuttle, J.P., pers. comm.).

manitobensis McDunnough - *Corylus* (McGugan *et al.*, 1958); *Quercus macrocarpa* (Ferguson, 1971-72); *Quercus* (Tietz, 1972).

oslari W. Rothschild - *Acer negundo, A. rubrum, A. saccharinum, Quercus velutina* (Tietz, 1972); *Quercus kelloggii* (Riotte & Peigler, 1980); *Quercus turbinella* (Lemaire, 1988); *Quercus arizonica* (Smith, M.J., pers. comm.); *Quercus gambelii* (Stone, S.E., pers. obs.); *Quercus oblongifolia, Q. palustris* (Tuttle, J.P., pers. comm.); *Quercus rubra* (Williams, B.D., pers. comm.).

peigleri Riotte - *Quercus palustris, Q. rubra, Q. marilandica, Q. velutina, Q. nigra, Q. falcata, Q. pubescens* (Riotte & Peigler, 1980); *Castanea vulgaris* (Lemaire, 1988); *Quercus gambelii* (Stone, S.E., pers. obs.).

pellucida discolor Ferguson - *Quercus nigra, Q. macrocarpa, Q. stellata, Q. velutina, Q. falcata, Q. marilandica* (Riotte & Peigler, 1980); *Quercus pubescens* (Lemaire, 1988); *Quercus palustris* (Tuttle, J.P., pers. comm.).

pellucida pellucida (J. E. Smith) - *Quercus nigra, Q. marilandica* (Ferguson, 1971-72); *Quercus falcata, Q. myrtifolia, Q. imbricaria, Q. nuttalli, Q. lyrata* (Riotte & Peigler, 1980); *Quercus pubescens, Q. ilex, Castanea vulgaris* (Lemaire, 1988); *Quercus velutina, Q. palustris* (Tuttle, J.P., pers. comm.).

senatoria (J. E. Smith) - *Fagus* (Forbes, 1923); *Quercus alba, Q. bicolor, Q. coccinea, Q. macrocarpa, Q. rubra, Q. velutina, Q. palustris* (McGugan *et al.*, 1958); *Carya, Castanea dentata, Corylus, Quercus ilicifolia, Q. imbricaria, Q. prinoides, Q. prinus, Rubus allegheniensis, ?R. idaeus* var. *strigosus* (Tietz, 1972); *Castanea vulgaris* (Lemaire, 1988).

---

[1] This was incorrectly reported as *Q. pini* in Riotte & Peigler (1980: 149) (Peigler, R. S., pers. comm.) and in Lemaire (1988: 312).

stigma (Fabricius) ssp. - *Quercus* (Morris, 1862); *Corylus* (Forbes, 1923); *Carya, Castanea dentata* (Tietz, 1972).

stigma fuscosa Ferguson - *Quercus nigra, Q. stellata, Q. falcata, Q. marilandica, Castanea vulgaris* (Riotte & Peigler, 1980); *Quercus palustris* (Tuttle, J.P., pers. comm.).

stigma stigma (Fabricius) - *Corylus* (Packard, 1905); *Quercus rubra* (McGugan *et al.*, 1958); *Quercus michauxii, Q. alba, Castanea mollissima* (Riotte & Peigler, 1980); *Quercus pubescens, Castanea vulgaris* (Lemaire, 1988); *Quercus nigra, Q. palustris, Q. velutina, Q. virginiana* (Tuttle, J.P., pers. comm.).

virginiensis (Drury) - *Quercus* (Holland, 1903); *Corylus* (Crotch, 1956); *Quercus macrocarpa* (McGugan *et al.*, 1958); *Quercus nigra* (Dominick & Edwards, 1971); *Carpinus caroliniana, Castanea dentata, Quercus alba, Q. prinus, Q. stellata* (Tietz, 1972); *Quercus rubra* (Riotte & Peigler, 1980); *Quercus robur, Q. petraea* (Naumann, S., pers. comm.); *Quercus velutina, Q. palustris* (Tuttle, J.P., pers. comm.).

ANTHERAEA Hübner

"Larvae of all but one species eat *Quercus* and also do well on *Fagus*. They also accept *Betula, Prunus, Salix, Carpinus, Castanea, Crataegus*" (Gardiner, 1982).

assamensis (Helfer) - *Quercus* (Crotch, 1956); *Cinnamomum obtusifolium, Laurus obtusifolia, Litsea citrata, L. polyantha, L. salicifolia, Machilus bombycina, M. odoratissima, Magnolia sphenocarpa, Michelia, Sarcostemma brevistigma, Symplocos grandiflora, Tetranthera diglottica, T. glauca, T. macrophylla, T. monopetala, T. polyantha* (Arora & Gupta, 1979); *Pyrus malus* (Albaugh, 1986); *Persea* (Holloway, 1987).

frithi (Moore) ssp. - *Shorea robusta* (Arora & Gupta, 1979).

frithi javanensis Bouvier - *Sapium sebiferum, Cinnamomum camphora* (Zhu & Wang, 1983).

godmani (Druce) - *Quercus* (Wolfe, K.L., pers. comm.).

harti Moore - *Quercus alba, Q. macrocarpa, Q. lyrata, Q. stellata, Q. palustris, Q. falcata, Q. phellos* (Furr, M., pers. comm.); *Quercus robur* (Naumann, S., pers. comm.).

korintjiana Bouvier - "Similar to ***Antheraea roylei***" (Holloway, 1987).

knyvetti Hampson - *Quercus, Prunus puddum* (Crotch, 1956); *Betula alnoides* (Arora & Gupta, 1979).

larissa Westwood - *Shorea glauca* (Holloway, 1987).

montezuma (Sallé) - *Quercus* (Smith, M.J., pers. comm.).

mylitta (Guérin-Méneville) - *Quercus, Fagus* (Crotch, 1956); *Quercus alba, Carpinus, Prunus* (Villiard, 1969); *Quercus robur* (Naumann, S., pers. comm.).

paphia Linnaeus - *Eugenia jambolana* (Watt, 1908); *Anogeissus latifolia, Bassia latifolia, B. longifolia, Bauhinia variegata, Bombax heptaphyllum, B. malabaricum, Canthium diccocum, Carpinus betulus, Careya arborea, C. sphaerica, Carissa carandas, Cassia lanceolata, Celastrus paniculatus, Chloroxylon swietenia,*

Cipadessa fruticosa, Dodonaea viscosa, Eugenia cuminii, Ficus benjamina, F. religiosa, F. retusa, Lagerstroemia indica, L. parviflora, Mangium caseolari, Pentaptera tomentosa, P. glabra, Prunus domestica, Rhizophora calceolaris, Ricinus communis, Shorea robusta, S. talura, Tectona grandis, Terminalia alata, T. arjuna, T. tomentosa, T. bellirica, T. catappa, Webera corymbosa, Ziziphus jujuba (Arora & Gupta, 1979); *Quercus, Salix* (Albaugh, 1986).

pernyi (Guérin-Méneville) - *Quercus, Betula, Carpinus, Prunus* (Crotch, 1956); *Quercus serrata* (Inoue *et al.*, 1982); *Fagus* (Rougeot & Viette, 1983); *Crataegus pinnatifida, Cinnamomum camphora, Xylosma japonicum* (Zhu & Wang, 1983); *Pyrus malus, Castanea, Salix* (Albaugh, 1986); *Quercus alba, Q. macrocarpa, Q. lyrata, Q. stellata, Q. palustris, Q. falcata, Q. phellos* (Furr, M., pers. comm.); *Quercus robur* (Naumann, S., pers. comm.).

polyphemus (Cramer) ssp. - *Quercus, Juglans cinerea, J. nigra, Carya, Tilia, Ulmus, Acer, Betula, Platanus, Castanea, Rosa, Fagus* (Holland, 1903); *Prunus, Corylus* (Saunders, 1909); *Tilia americana, T. europaea, Prunus virginiana, Hamamelis virginiana, Castanea vesca, Crataegus tomentosa, C. coccinea, C. crus-galli, Amelanchier canadensis, Ribes cynosbati, Corylus americana, C. rostrata, Fagus ferruginea, Carpinus americana, Ostrya virginiana, Carya tomentosa, C. amara, Betula lenta, B. alba, Alnus incana, A. serrulata, Salix, Salix alba, S. humilis, Populus grandidentata, P. tremuloides, Baptisia tinctoria, Vaccinium, Gleditsia, Quercus alba, Q. nigra, Q. macrocarpa, Q. rubra, Q. virens* (Packard, 1914); *Cornus* (Galtsoff, *et al.*, 1937); *Sorbus, Weigela,* "rosaceous trees" (Ferguson, 1971-72); *Fraxinus* (Baker, 1972); *Myrica cerifera* (Brown, 1972); *Acer pensylvanicum, A. rubrum, A. saccharinum, Alnus, Arbutus menziesii, Capsicum annuum, Carpinus caroliniana, Castanea dentata, C. sativa, Cydonia oblonga, Gaylussacia dumosa, Gleditsia triacanthos, Liriodendron tulipifera, Malus pumila, Pinus, Platanus occidentalis, Prunus cerasus, P. domestica, P. domestica* var. *galatensis, P. persica, P. serotina, Pyrus communis, Quercus coccinea, Ribes sanguineum, R. sativum, Sassafras albidum, Ulmus, Vitis* (Tietz, 1972); *Betula pendula, Carya glabra, C. ovata, Cornus racemosa, Quercus* X *bebbiana, Q. velutina, Salix babylonica, S. fragilis* (Wagner & Mayfield, 1980); *Quercus lyrata, Q. stellata, Q. falcata, Q. phellos* (Furr, M., pers. comm); *Quercus robur* (Naumann, S., pers. comm.); *Betula populifolia* (Sblendorio, D.C., pers. comm.); *Rhamnus, Sambucus* (Stone, S.E., pers. obs.); *Quercus palustris, Salix interior, Ulmus americana, Acer negundo* (Tuttle, J.P., pers. comm.); *Betula papyrifera* (Williams, B.D., pers. comm.).

roylei Moore - *Quercus* (Crotch, 1956); *Bassia latifolia, B. longifolia, Betula alnoides, Cyperus esculentus, Daphniphyllum himalense, Euodia fraxinifolia, Prunus puddum, Pyrus communis, P. malus, Quercus dealbata, Q. dilatata, Q. glauca, Q. himalayana, Q. ilex, Q. incana, Q. semicarpifolia, Q. semiserrata, Q. serrata* (Arora & Gupta, 1979); *Quercus robur, Q. turneri* (Naumann, S., pers. comm.).

semperi Felder - *Quercus turneri* (Naumann, S., pers. comm.).

yamamai (Guérin-Méneville) ssp. - ***Quercus, Castanea*** (Crotch, 1956); *Castanea crenata, **Quercus acutissima, Q. dentata, Q. serrata, Malus pumila**, Prunus* (Inoue *et al.*, 1982); ***Pyrus malus**, Fagus, Carya, **Morus*** (Albaugh, 1986); *Quercus robur* (Naumann, S., pers. comm.).
yamamai superba Inoue - *Quercus acutissima* (Wang, 1988).

ANTHERINA Sonthonnax
suraka (Boisduval) - *Nerium* (Bouvier, 1936); *Vitis vinifera, Schinus, Eugenia, Maesa, Strophanthus*, "fruit trees" (Griveaud, 1961); *Ligustrum japonicum, Liquidambar, Salix, Fagus, Crataegus* (Probst, H., pers. comm.).

ANTISTATHMOPTERA Tams
daltonae Tams - ?*Cordyla africana* (Gardiner, 1982).

ARCHAEOATTACUS J. H. Watson
edwardsii (White) - *Ligustrum, Syringa vulgaris, Salix, Prunus, Populus* (Villiard, 1969); "foodplant preferences of the larvae coincide with those of *Attacus atlas* and include *Ailanthus altissima* and especially LAURACEAE" (Peigler, 1989); *Ilex chinensis* (Zhu & Wang, 1983).

ARGEMA Wallengren
besanti Rebel - *Sclerocarya caffra, Spirostachys africana, Commiphora mollis, Juglans regia, Eucalyptus, Liquidambar* (Gardiner, 1982).
kuhnei Pinhey - *Monotes katangensis* (Pinhey, 1972).
mimosae (Boisduval) - *Sclerocarya caffra* (Platt, 1921); *Juglans regia* (Crotch, 1956); *Commiphora mollis, Spirostachys africana* (Pinhey, 1972); *Eucalyptus, Liquidambar* (Gardiner, 1982); *Sclerocarya birrea* (Oberprieler, 1987); *Rhus laurina* (Wolfe, K.L., pers. comm.).
mittrei (Guérin-Méneville) - *Rhus toxicodendron, Schinus molle, Mimosa, Eugenia cuneifolia, Weinmannia eriorcampa* (Villiard, 1969); *Uapaca*, MIMOSACEAE (=FABACEAE: MIMOSOIDEAE) (Gardiner, 1982); *Rhus copallina, R. typhina* (Albaugh, 1986); *Rhus laurina* (Wolfe, K.L., pers. comm.).

ARSENURA Duncan
Arsenura sp. - ?MALVACEAE, ?TILIACEAE (Crotch, 1956); *Chorisia* (Gardiner, 1982).
armida (Cramer) ssp. - "Palisade tree", *Annona, Bombax ceiba* (Packard, 1914); *Bombacopsis quinatum* (Janzen, 1984a); *Chorisia insignis, C. speciosa* (Wolfe, K.L., pers. comm.).
aspasia (Herrich-Schäffer) - "A Melastomaceous tree" (Packard, 1914).
polyodonta (Jordan) - *Chorisia insignis* (Gardiner, 1982); *Chorisia speciosa* (Wolfe, K.L., pers. comm.).
rebeli Gschwandner - *Tilia* (Herbin, D., pers. comm.).
xanthopus (Walker) - *Urena sinuata* (Packard, 1914).

ATHLETES Karsch

gigas Sonthonnax - ***Brachystegia randii*** (Bouvier, 1936); ?*Cercis siliquastrum*, "?woody" LEGUMINOSAE (=FABACEAE) (Crotch, 1956); *Julbernardia paniculata* (Pinhey, 1972).

semialba Santhonnax - *Brachystegia randii, B. globifera, Cassia* (Platt, 1921); *Brachystegia spiciformis* (Pinhey, 1972).

ATTACUS Linnaeus

atlas (Linnaeus) - *Phyllanthus emblica* (Moore, 1859); *Glochidion velutinum* (=*Bradleia ovata* & *Phyllanthus velutinus*), *Sapium insigne* (Moore, 1862); *Pyrus malus, P. communis* (Wailly, 1879); *Prunus* (Poujade, 1880); *Schleichera oleosa, Nauclea rotundifolia, Leucosceptrum canum* (=*Teucrium macrostachyum*), *Lagerstroemia indica* (Cotes, 1891-93); *Nerium oleander, Quisqualis indica, Sapium sebiferum* (Fauvel, 1895); *Artemisia vulgaris, Symplocos paniculata* (=*S. crategoides*) (Silbermann, 1897); *Ailanthus altissima* (Voelschow, 1902); *Cinchona succirubra, Coffea arabica* (Fruhstorfer, 1904); *Clerodendrum serratum, C. viscosum* (=*C. infortunatum*), *Dillenia pentagyna, Melastoma malabathricum, Meyna grisea* (=*Vangueria spinosa*), *Morus* (André, 1908); "Samalu hedges" (Packard, 1914); *Dillenia indica* (Rama Rao, 1914); *Camellia* (=*Thea*) *sinensis, Ceiba pentandra, Hibiscus, Piper, Stachytarpheta cayennensis, Theobroma cacao* (Jurriaanse & Lindemans, 1920); CUPULIFERAE (archaic group name for *Quercus, Carpinus*, etc.) (Seitz, 1926); *Aleurites montana, Canarium indicum* (=*C. commune*), *Uncaria gambir* (Dammerman, 1929); *Alstonia scholaris, Cinnamomum iners, Clidemia hirta, Curcuma viridiflora* (=*C. longa*), *Ficus, Nephelium lappaceum, Lantana camara, Lannea grandis, Lonicera japonica, Nephelium lappaceum, Nicolaia speciosa, Parkia intermedia, P. speciosa, Ricinus communis, Rosa, Sesbania grandiflora, Spathodea campanulata, Syzygium aqueum, S. malaccense* (=*E. malaccensis*) (Dupont & Scheepmaker, 1936); *Fraxinus, Kalmia latifolia, Rhododendron, Sassafras albidum, Syringa vulgaris* (Villiard, 1969); *Euphorbia longana, Litchi chinensis* (=*Nephelium litchi*) (SsuMa, 1972); *Populus* (Dickens, 1973); *Schefflera octophylla* (Hill & Cheung, 1978); *Carpinus betulus, Sarcostemma brunonianum* (=*S. brevistigma*) (Arora & Gupta, 1979); *Ardisia sieboldii* (Inoue *et al.*, 1982); *Betula platyphylla, Ilex chinensis, Ipomoea batatas, Setaria viridis* (Zhu & Wang, 1983); *Glochidion rubrum* (Wang, 1988); *Annona muricata, Averrhoa carambola, Berberis asiatica, B. vulgaris, B. thunbergii, Bischofia javanica, Canangium odoratum* (=*Cananga odorata*), *Cinnamomum camphora, Cinnamomum zeylanicum, Citrus, Erythrina subumbrana, Ligustrum, Mangifera indica, Persea americana, P. pyriformis* (=*P. gratissima*), *Salix, Sandoricum koetjape, Swietenia mahogani* (Peigler, 1989).

caesar Maassen - *Syringa vulgaris, Annona squamosa, Artocarpus integer, Canangium odoratum* (=*Cananga odorata*), *Citrus, Erythrina subumbrana, Myrica cerifera, Persea americana, P. pyriformis* (=*P. gratissima*), *Salix, Sandoricum koetjape, Swietenia mahogany* (Peigler, 1989).

erebus Fruhstorfer - ?*Ficus* (Ver Huell, 1836).

lorquinii C. & R. Felder - *Psidium guajava* (=*P. pyriforme*), *P. cujavillus* (Laporte, 1830); *Quercus* (Gosse, 1879); *Theobroma cacao* (Jurriaanse & Lindemans, 1920); *Fraxinus, Syringa vulgaris, Annona squamosa, Artocarpus integra, Canangium odoratum* (=*Cananga odorata*), *Chrysophyllum, Citrus, Heliconia, Ligustrum, Magnolia virginiana, Mangifera indica, Myrica cerifera, Persea americana, P. pyriformis* (=*P. gratissima*), *Persea borbonia, Quercus glauca* (=*Cyclobalanopsis glauca*), *Salix, Sandoricum koetjape, Schefflera arboricola* (=*Brassaia actinophylla* var. *arboricola*) (Peigler, 1989).

mcmulleni J. H. Watson - "Samalu hedges" (Packard, 1914).

taprobanis Moore - ?*Ocimum* (Cotes, 1891-93); *Aglaia roxburghiana* (=*Milnea roxburghiana*) (Voelschow, 1902); *Berberis asiatica, B. vulgaris, B. thunbergii, Cinnamomum camphora, Cinnamomum zeylanicum, Ligustrum* (Peigler, 1989).

AURIVILLIUS Packard

arata (Westwood) - *Albizia fastigiata* (Platt, 1921); *Piptadenia buchanani* (Rougeot, 1962); *Albizia gummifera, Parkia biglobosa, Peltophorum africanum, Piptadenia* (Pinhey, 1972); *Salix* (Gardiner, 1982).

triramis Rothschild - *Albizia* (Rougeot, 1962).

AUTOMERELLA Michener

flexuosa Felder - *Fagus, Quercus, Robinia pseudoacacia, Laburnum* (Gardiner, 1982).

AUTOMERINA Michener

cypria (Gmelin) - *Theobroma cacao* (Gardiner, 1982).

vala (Kirby) - *Theobroma cacao* (Gardiner, 1982).

AUTOMERIS Hübner

Automeris sp. - *Brosimun utile, Cordia alliodora, Juglans neotropica, Cupressus, Schizolobium parahybum, Swietenia macrophylla, Virola sebifera, Prunus capuli, Cecropia, Leucaena leucocephala* (Gara & Onore, 1989).

amanda Schaus ssp. - *Piptadenia, Selinum, Gourliea* (=*Geoffroea*) *decorticans*, LEGUMINOSAE (=FABACEAE) (Gardiner, 1982).

aurantiaca Weymer - See ***Automeris naranja*** Schaus ssp. (Lemaire, 1971-74: 300).

banus (Boisduval) ssp. - *Gleditsia triacanthos* (Herbin, D., pers. comm.); *Robinia pseudoacacia, Rhus laurina, Pyrus malus* (Wolfe, K.L., pers. comm.).

beckeri (Herrich-Schäffer) - *Sorocea ilicifolia, Fagus, Prunus, Quercus*, URTICACEAE (Gardiner, 1982).

boucardi Druce - *Laburnum* (Gardiner, 1982).

boudinotiana Lemaire - *Quercus* (Wolfe, K.L., pers. comm.).

cecrops cecrops (Boisduval) - *Liriodendron tulipifera* (Lemaire, C., pers. comm.).

cecrops peigleri Lemaire - *Quercus grisea* (Lemaire, 1979b).

colenon Dyar - "Grasses, esp." *Cynodon* (Smith, M.J., pers. comm.).

complicata (Walker) - *Pyrus communis, Bauhinia forficata, Tipuana speciosa, Wisteria sinensis, Prunus triflora, Lagerstroemia indica, Spartium junceum, Psidium guajava, Citrus aurantium, Passiflora* (Gardiner, 1982).

coresus (Boisduval) - *Platanus orientalis, Acer, Ligustrum* (Crotch, 1956); *Quercus robur* (Naumann, S., pers. comm.).

excreta Draudt - *Robinia pseudoacacia, Prunus, Pyrus malus* (Wolfe, K.L., pers. comm.).

godartii (Boisduval) - *Gleditsia triacanthos* (Herbin, D., pers. comm.).

hamata Shaus - *Robinia pseudoacacia, Laburnum, Fagus* (Gardiner, 1982); *Salix* (Herbin, D., pers. comm.).

illustris (Walker) - *Platanus orientalis, Ficus benjamina, Ligustrum, Syringa vulgaris, Fagus, Crataegus* (Gardiner, 1982).

incarnata (Walker) - *Eucalyptus, Robinia pseudoacacia, Laburnum, Fagus* (Gardiner, 1982).

inornata (Walker) - *Robinia pseudoacacia, Prunus, Quercus* (Gardiner, 1982).

io (Fabricius) ssp. - *Laurus sassafras* (=*Sassafras albidum*), *Cornus florida, Liriodendron* (Morris, 1862); *Pyrus malus, Prunus, Salix, Ulmus, Populus balsamifera, Quercus, Ribes, Melilotus, Gossypium herbaceum, Gleditsia triacanthos*, "thorn" (=*Crataegus*) (Saunders, 1909); *Baptisia tinctoria* (Packard, 1914); *Liquidambar styraciflua, Nephelium litchi, Persea americana* (Brown, 1972); *Abies balsamea, Acer, Amorpha fruticosa, Betula, Carpinus caroliniana, Carya, Cassia fasciculata, Cercis canadensis, Comptonia peregrina, Corylus, C. avellana, Fagus, Fraxinus, Hibiscus lasiocarpos, Humulus lupulus, Ilex verticillata, Lespedeza, Liriodendron tulipifera, Magnolia virginiana, Malus pumila, Melilotus alba, M. officinalis, Myrica pensylvanica, Ostrya virginiana, Platanus occidentalis, Populus gileadensis, P. tremuloides, Prunus cerasus, P. domestica, P. pensylvanica, P. serotina, P. virginiana, Pyrus communis, Rhizophora mangle, Rhododendron, Ribes sativum, Robinia pseudoacacia, Rosa, Rubus allegheniensis, R. canadensis, R. idaeus* var. *strigosus, Sassafras albidum* var. *molle, Serenoa repens, Spirea, Symphoricarpos albus, Tilia, Trifolium, T. repens, Viburnum dentatum, Wisteria frutescens, W. sinensis, Zea mays* (Tietz, 1972); *Celtis laevigata, Malvaviscus arboreus* var. *drummondii, Leucaena pulverulenta* (Neck, 1976); *Populus deltoides, Quercus palustris, Salix interior, Acer rubrum* (Tuttle, J.P., pers. comm.); *Mimosa* (Sblendorio, D.C., pers. comm.); *Robinia* (Wolfe, K.L., pers. comm.).

io coloradensis Cockerell - *Fagus* (Probst, H., pers. comm.); *Glycyrrhiza, Pyrus malus, Quercus gambelii, Salix, Tilia americana* (Stone, S.E., pers. obs.).

io draudtiana Lemaire - *Gleditsia* (Smith, M.J., pers. comm.); *Robinia pseudoacacia* (Wolfe, K.L., pers. comm.).

io io (Fabricius) - *Sassafras, Cornus, Zea mays* (Smith, 1797); *Gossypium herbaceum, Fagus* (Packard, 1914); *Myrica carolinensis, Salix, Populus, Carya, Fraxinus, Celtis pallida, Ulmus, Robinia pseudoacacia, Quercus, Tilia, Trifolium, Baptisia tinctoria, Cassia chamaecrista*, "bean", *Amorpha, Betula, Phaseolus vulgaris,* "rosaceous trees and shrubs" (Ferguson, 1971-72); *Populus, Morus, Platanus* (Baker, 1972); *Crescentia alata, Mimosa pigra, Cassia biflora, Rhynchosia reticulata, Gliricidia sepium* (Janzen, 1984a); *Syringa vulgaris, Sassafras albidum, Acer negundo, Rosa, Liriodendron tulipifera, Juglans regia, Liquidambar styraciflua, Diospyros, Ligustrum, Crataegus* (Albaugh, 1986); *Gleditsia triacanthos* (Curtis, W.M., pers. comm.); *Prunus serotina* (Williams, B.D., pers. comm.).

io lilith (Strecker) - *Serenoa repens* (Riley, 1889); *Rhizophora* (Packard, 1893); *Conocarpus erectus* (Packard, 1914); *Persea americana, Rosa, Cercis canadensis, Rhododendron, Zea mays, Gossypium* (probably *G. hirsutum*), *Litchi chinensis, Serenoa repens, Hibiscus, Rhapis, Turnera ulmifolia, Tabebuia argentea, Galactia, Amorpha fruticosa* (Kimball, 1965); *Prunus virginiana* (Sblendorio, D.C., pers. comm.); *Prunus serotina* (Williams, B.D., pers. comm.).

io neomexicana Barnes & Benjamin - *Salix capraea* (Naumann, S., pers. comm.); *Salix, Glycyrrhiza* (Stone, S.E., pers. obs.); *Quercus rubra* (Williams, B.D., pers. comm.).

iris hesselorum Ferguson - *Chilopsis linearis, Eysenhardtia polystachya* (Tuskes, 1985b); *Quercus, Mimosa* (Smith, M.J., pers. comm.); *Quercus palustris* (Tuttle, J.P., pers. comm.); *Robinia pseudoacacia* (Wolfe, K.L., pers. comm.).

iris iris (Walker) - *Robinia* (Gardiner, 1982).

janus (Cramer) - *Erythrina, Fagus, Pyrus malus* (Gardiner, 1982).

jucunda (Cramer) - *Robinia pseudoacacia, Fagus, Prunus* (Gardiner, 1982).

leucane (Geyer) - See ***Leucanella leucane*** (Geyer) (Lemaire, 1971-74: 383).

liberia (Cramer) - *Citrus, Platanus* (Gardiner, 1982); *Gleditsia triacanthos* (Lemaire, C., pers. comm.); *Robinia pseudoacacia, Pyrus malus* (Wolfe, K.L., pers. comm.).

louisiana Ferguson & Brou - *Prunus serotina, Quercus virginiana* (Ferguson & Brou, 1981); *Salix babylonica* (Tuttle, J.P., pers. comm.).

melanops (Walker) - *Terminalia catappa, Hibiscus tiliaceus, Tamarindus indica, Schinus terebinthifolius, Platanus orientalis, Tipuana speciosa* (Gardiner, 1982).

metzli (Sallé) - *Platanus occidentalis, Erythrina rubra, Crataegus* (Gardiner, 1982); *Quercus alba, Q. coccinea* (Purdum, D.A., pers. comm.); *Pyrus malus* (Smith, M.J., pers. comm.).

montezuma (Boisduval) - *Robinia pseudoacacia* (Wolfe, K.L., pers. comm.).

moresca Schaus - *Robinia pseudoacacia* (Wolfe, K.L., pers. comm.).

muscula Vuillot - *Rosa* (Crotch, 1956).

naranja Schaus ssp. - *Ligustrum* (Crotch, 1956); *Syringa vulgaris, Salix, Prunus* (Villiard, 1969); *Quercus* (Gardiner, 1982).

oberthurii (Boisduval) - *Prosopis alba, P. nigra* (Gardiner, 1982).

pamina (Neumoegen) - *Quercus undulata, Q. emoryi, Cercocarpus parvifolius, Ceanothus fendleri, Opuntia* (Kunzé, 1900); *Cercocarpus betuloides* var. *blancheri, Quercus marilandica* (Tietz, 1972); *Robinia neomexicana* (McFarland, 1975); *Celtis pallida* (Smith, M.J., pers. comm.); *Quercus gambelii* (Stone, S.E., pers.

obs.); ***Quercus palustris*** (Tuttle, J.P., pers. comm.); ***Rhus laurina*** (Wolfe, K.L., pers. comm.); ***Quercus rubra*** (Williams, B.D., pers. comm.).

phrynon Druce - ***Rhus laurina*** (Wolfe, K.L., pers. comm.).

randa Druce - ***Quercus, Celtis pallida*** (Tuskes, 1985b); ***Quercus*** *gambelii* (Stone, S.E., pers. obs.); ***Prunus serotina, Quercus palustris*** (Tuttle, J.P., pers. comm.); ***Quercus*** *rubra* (Williams, B.D., pers. comm.).

rubrescens (Walker) ssp. - ***Sida vera*** (Janzen, 1982); ***Inga vera,*** *Rourea glabra, Guazuma ulmifolia, Cassia biflora, C. grandis, Quercus oleoides, Cordia alliodora, Lonchocarpus minimiflorus, Calycophyllum candidissimum, Crescentia alata, Zuelania guidonia* (Janzen, 1984a); ***Quercus robur*** (Naumann, S., pers. comm.); *Salix, Syringa vulgaris* (Simon, L.D., pers. comm.).

saturata (Walker) - See ***Leucanella saturata*** (Walker) (Lemaire, 1971-74: 411).

tridens (Herrich-Schäffer) - ***Ligustrum*** (Herbin, D., pers. comm.); ***Quercus*** (Purdum, D.A., pers. comm.).

vinosus Conte - See ***Automeris incarnata*** (Walker) (Lemaire, 1971-74: 244).

zephyria Grote - ***Quercus gambelii*** (Tietz, 1972); ***Prunus*** *serotina* (Donahue, 1979); *Salix, Cercocarpus betuloides, Cercis canadensis, Quercus phellos, Q. nigra, Q. oblongifolia, Q. alba* (Tuskes & Smith, 1988); ***Quercus*** *robur, Q. petraea* (Naumann, S., pers. comm.); ***Tilia, Prunus virginiana*** (Stone, S.E., pers. obs.); ***Quercus palustris, Salix babylonica*** (Tuttle, J.P., pers. comm.); ***Quercus*** *rubra* (Williams, B.D., pers. comm.); ***Pyrus malus*** (Wolfe, K.L., pers. comm.).

zozine Druce - ***Quercus robur*** (Naumann, S., pers. comm.); ***Rhus*** *laurina, Robinia pseudoacacia* (Wolfe, K.L., pers. comm.).

zugana Druce - *Annona purpurea, Lonchocarpus costaricensis, L. eriocarinalis, Quercus oleoides, Cydista heterophylla, Calycophyllum candidissimum, Hymenaea courbaril, Solanum hazenii, Lantana camara, Centrosema pubescens, Cassia hayesina, Inga vera, Serjania atrolineata* (Janzen, 1984a); *Ulmus* (Ringle, G., pers. comm.).

BUNAEA Hübner

alcinoe (Stoll) - ***Cussonia spicata, Ekebergia meyeri*** (Platt, 1921); *Maesa lanceolata* (Bouvier, 1936); ***Prunus, Crataegus*** (Crotch, 1956); *Harpephyllum caffrum* (Taylor, 1957); ***Khaya senegalensis, Terminalia catappa*** (Rougeot, 1962); ***Rhus*** *glabra, Rosa, Pyrus malus* (Villiard, 1969); ***Bauhinia reticulata,*** *Celtis, Croton, E. ruepellii, Gymnosporia senegalensis* (Pinhey, 1972); ***Ligustrum*** (Gardiner, 1982); ***Quercus robur*** (Naumann, S., pers. comm.).

heroum Oberthür - See ***Imbrasia macrothyris*** Rothschild (Pinhey, 1972: 94).

BUNAEOPSIS Bouvier

arabella (Aurivillius) - "A reedlike grass" (Platt, 1921).

aurantiaca (Rothschild) - "Grasses", ***Bridelia micrantha*** (Pinhey, 1972).

bomfordi Pinhey - *Dolichos* (Pinhey, 1972).

hersilia (Westwood) - GRAMINEAE (=POACEAE) (Crotch, 1956); *Chrysophyllum* (Rougeot, 1962); *Hyparrhenia contracta, Bequaertiodendron, Irvingia, Pennisetum* (Pinhey, 1972); "several genera of bamboo" (Gardiner, 1982).
hersilioides Fleury - *Annona senegalensis* (Bouvier, 1936).
jacksoni (Jordan) - *Scleria rehmanii* "and a reedlike grass" (Pinhey, 1972).
licharbas (Maassen & Weymer) - "Leguminous plants" (Pinhey, 1972); *Quercus, Salix capraea* (Gardiner, 1982).
oubie (Guérin-Méneville) - "Grasses" (Platt, 1921); *Pycreus muricatus* (Crotch, 1956); *Euclea* (Pinhey, 1956); *Fagus, Salix* (Gardiner, 1982).
phidias (Weymer) - "?Grasses" (Pinhey, 1972).
zaddachi (Dewitz) - "Grasses", *Euclea, Sporobolus* (Pinhey, 1972); *Heteropogon* (Pinhey, 1975).

CAIO Travassos & Noronha
championi (Druce) - *Bombacopsis quinatum* (Janzen, 1984a); *Chorisia, Tilia* (Wolfe, K.L., pers. comm.).
richardsoni (Druce) - *Chorisia, Ceiba acuminata* (Smith, M.J., pers. comm.).

CALIGULA Moore
boisduvali (Eversmann) ssp. - *Quercus, Castanea, Acer, Zelkova serrata, Acacia, Juglans, Prunus, Pyrus, P. communis, Sorbus* (Villiard, 1969); *Lonicera japonica, Cornus controversa, Fagus crenata, Rosa* (Inoue *et al.*, 1982).
boisduvali fallax (Jordan) - *Corylus heterophylla, Tilia, Quercus* (Zhu & Wang, 1983).
cachara Moore - *Crataegus, Quercus acuta, Q. ilex, Juglans regia* (Crotch, 1956).
japonica Moore ssp. - *Castanea crenata, Prunus, Pyrus malus, Quercus, Salix, Crataegus, Viburnum opulus* var. *sterile* (Crotch, 1956); *Schleichera trijugata, Rhus verniciflua, Zelkova serrata, Acer, Prunus persica, P. armeniaca, Cinnamomum camphora, Lagerstroemia indica, Platanus, Cercis siliquastrum* (Villiard, 1969); *Juglans regia* (Gardiner, 1982); *Castanea mollissima, Prunus salicina, Pyrus communis, Corylus heterophylla, Catalpa bungei* (Zhu & Wang, 1983); *Aesculus hippocastanum* (Naumann, S., pers. comm.); *Populus deltoides* (Peigler, R.S., pers. comm.); *Prunus virginiana* (Stone, S.E., pers. obs.).
japonica arisana Shiraki - *Castanea crenata, Cinnamomum camphora, Juglans cathayensis, Quercus acutissima, Ginkgo biloba* (Wang, 1988).
jonasi Butler - *Populus deltoides* (Peigler, R.S., pers. comm.).
simla (Westwood) - *Salix fragilis* (Packard, 1914); *Salix babylonica, Pyrus communis, P. malus, Crataegus* (Crotch, 1956); *Betula, Quercus, Carya* (Villiard, 1969); *Juglans regia* (Gardiner, 1982); *Crataegus oxyacantha* (Naumann, S., pers. comm.).
thibeta (Westwood) ssp. - *Andromeda ovalifolia* (Bouvier, 1936).
thibeta extensa (Butler) - *Ligustrum ovalifolium* (Nardelli, 1986).
thibeta okurai Okano - *Glochidion hongkongense, Quercus variabilis* (Wang, 1988).
zuleika (Westwood) - *Acer caudatum, Actinodaphne sikkimensis* (Bouvier, 1936).

CALLOSAMIA Packard

angulifera (Walker) - ***Liriodendron tulipifera*** (Holland, 1903); *Sassafras albidum* (Ferguson, 1971-72); ***Prunus serotina*** (Tietz, 1972); ***Magnolia tripetala*** (Peigler, 1988).

promethea (Drury) - ***Laurus sassafras*** (=***Sassafras albidum***), ***L. benzoin***, *Cephalanthus occidentalis* (Morris, 1862); ***Liquidambar styraciflua***, *Liriodendron tulipifera*, ***Malus pumila***, ***Populus***, ***Pyrus communis***, ***Prunus***, ***P. domestica***, ***Salix***, ***Syringa vulgaris***, ***Viburnum*** (Eliot & Soule, 1902); "a predilection for the LAURACEAE" (Holland, 1903); ***Fraxinus***, ***Rhododendron*** (Saunders, 1909); *Prunus serotina*, ***Berberis***, ***Betula*** (Packard, 1914); ***Fraxinus pennsylvanica***, ***Prunus pensylvanica*** (McGugan *et al.*, 1958); ***Acer***, ***Berberis vulgaris***, ***Fraxinus americana***, *Tilia* (Collins & Weast, 1961); ***Ligustrum*** (Villiard, 1969); *Diospyros*, *Symplocos tinctoria*, *Corylus*, ***Prunus virginiana*** (Ferguson, 1971-72); *Fagus*, *Halesia carolina*, ***Magnolia***, ***Myrica pensylvanica***, ***Pinus***, ***Prunus persica***, *Quercus*, *Styrax americana*, ***Thuja occidentalis***, ***Tilia americana*** (Tietz, 1972); *Magnolia virginiana* (Peigler, 1976); ***Betula papyrifera*** (Gardiner, 1982); ***Kalmia*** (Albaugh, 1986); ***Magnolia acuminata***, ***Cinnamomum camphora***, ***Gordonia lasianthus*** (Peigler, 1988).

securifera (Maassen) - ***Magnolia virginiana*** (Jones, 1909); *Gordonia lasianthus*, *Magnolia grandiflora*, ***Liriodendron tulipifera*** (Peigler, 1988); *Cinnamomum camphora* (Simon, L.D., pers. comm.).

CARNEGIA Aurivillius

mirabilis Aurivillius - ***Annona senegalensis*** (Rougeot, 1962).

CATOCEPHALA Blanchard

socialis Feisthamel - *Cryptocarya peumus* (Gardiner, 1982).

CICIA Oiticica

crocata (Boisduval) - *Gleditsia triacanthos* (Packard, 1905).

CINABRA Sonthonnax

hyperbius (Westwood) - ***Brachystegia randii***, ***Protea*** (Platt, 1921); ***Berlinia paniculata***, ***Brachystegia venosa*** (Crotch, 1956); ***Brachystegia spiciformis***, *Julbernardia paniculata* (Pinhey, 1972); ***Ekebergia*** (Pinhey, 1975); ***Prunus***, *Robinia pseudoacacia* (Gardiner, 1982).

CIRINA Walker

forda (Westwood) - ***Sideroxylon inerme*** (Dickerson, 1945); ***Rhus*** *longifolia*, *Carissa grandiflora* (Platt, 1921); ***Prunus***, ***Pyrus malus*** (Crotch, 1956); *Hippobromus pauciflora*, ***Pappea capensis***, ***Euclea undulata***, ***Schotia speciosa*** (Taylor, 1957); ***Erythrophleum guineense***, ***Faidherbia albida*** (Rougeot, 1962); "woody shrubs," ***Albizia versicolor***, ***Carissa edulis***, ***Ekebergia meyeri***, ***Schinus molle***, *Warburgia*

(Pinhey, 1972); *Aloe* (Pinhey, 1975); *Acacia nilotica, Burkea africana* (Gardiner, 1982); *Rhus laurina* (Wolfe, K.L., pers. comm.).

CITHERONIA Hübner

andina Lemaire - *Schinus terebinthifolius* (Wolfe, K.L., pers. comm.).

armata armata W. Rothschild - *Anacardium occidentale* (Lemaire, 1988).

azteca Schaus ssp. - *Rhus typhina* (Purdum, D.A., pers. comm.); *Rhus laurina, Schinus terebinthifolius* (Wolfe, K.L., pers. comm.).

beledonon Dyar - *Rhus typhina* (Purdum, D.A., pers. comm.); *Liquidambar styraciflua, Ligustrum* (Simon, L.D., pers. comm.); *Gossypium* (Smith, M.J., pers. comm.); *Rhus laurina, Schinus terebinthifolius* (Wolfe, K.L., pers. comm.).

bellavista Draudt ssp. - *Phoradendron quadrangulare* (Janzen, 1984a); *Rhus laurina* (Wolfe, K.L., pers. comm.).

brissotii (Boisduval) ssp. - *Excoecaria biglandulosa* (Packard, 1905); *Ligustrum*, EUPHORBIACEAE (Crotch, 1956); *Juglans, Carya illinoensis, Carya* (Albaugh, 1986); *Prunus, Amygdalus communis* (Lemaire, 1988).

brissotii meridionalis Bouvier - *Prunus persica* (Dias, 1978).

hamifera W. Rothschild ssp. - *Terminalia catappa, Salix capraea, Ligustrum* (Lemaire, 1988).

laocoon (Cramer) - *Aleurites fordii, Psidium albortum, P. araca, P. guajava, P. incanescens, P. mediterraneum, P. petrosum, P. pomiferum, Acalypha, Caryocar brasiliensis, Terminalia catappa, Rosa, Joannesia princeps, Anacardium occidentale, Gossypium herbaceum, Ilex paraguaiensis, Zinnia elegans, Schinus terebinthifolius, Sebastiania klotzschiana, Vernonia polyanthus, V. tweedieana, Diospyros kaki, Cupressus* (Dias, 1978); COMBRETACEAE, ROSACEAE, MYRTACEAE, CARYOCARACEAE, MALVACEAE, ANACARDIACEAE, AQUIFOLIACEAE, *Ricinus communis* (Lemaire, 1988).

lobesis W. Rothschild ssp. - *Cochlospermum vitifolium, Bursera simaruba, Spondias mombin, Psidium guajava, Calycophyllum candidissimum, Phoradendron quadrangulare* (Janzen, 1984a).

mexicana Grote & Robinson - *Juglans, Carya* (Albaugh, 1986).

phoronea (Cramer) - "Fruta de erbe" (Packard, 1905); *Bixa orellana* (Dias, 1978); *Urucu* (=*Bixa*) (Lemaire, 1988).

regalis (Fabricius) - *Juglans regia, Diospyros virginiana* (Morris, 1862); *Juglans cinerea, Rhus* (Holland, 1903); *Gossypium herbaceum, Juglans nigra* (Packard, 1905); *Liquidambar styraciflua* (Forbes, 1923); *Corylus, Ligustrum* (Crotch, 1956); *Carya* (English, 1958); *Rhus glabra, R. copallina, R. cismontana, Salix, Quercus* (Villiard, 1969); *Cephalanthus occidentalis, Fraxinus, Oxydendrum arboreum, Platanus occidentalis, P. orientalis, Prunus domestica, Sassafras albidum* var. *molle* (Tietz, 1972); *Nyssa sylvatica, Syringa vulgaris* (Ferguson, 1971-72); *Carya illinoensis, Diervilla* (Albaugh, 1986); *Rhus typhina* (Lemaire, 1988); *Carya glabra, Prunus serotina* (Tuttle, J.P., pers. comm.); *Carya ovata* (Williams, B.D., pers. comm.); *Rhus laurina* (Wolfe, K.L., pers. comm.).

sepulcralis Grote & Robinson - ***Pinus*** (Holland, 1903); ***Pinus caribaea*** (Kimball, 1965); ***Pinus rigida, P. strobus*** (Packard, 1905); ***Pinus nigra, P. mitis, P. sylvestris*** (Crotch, 1956); ***Pinus halepensis*** (Lemaire, 1988); ***Pinus mugo*** (Naumann, S., pers. comm.); ***Pinus ponderosa*** (Stone, S.E., pers. obs.).

splendens (Druce) ssp. - ***Lithrea brasiliensis, Schinus dependens, Loranthus flexicaulis, Mangifera indica*** (Dias, 1978); ***Rhus*** (Platt & Williams, 1982).

splendens sinaloensis C. C. Hoffmann - ***Gossypium thurberi*** (Weast, 1962); ***Arctostaphylos pungens, Rhus choriophylla*** (Tuskes, 1986b); ***Juglans nigra, Pistacia lentiscus*** (Lemaire, 1988); ***Arbutus menziesii, Juglans regia, Liquidambar styraciflua, Schinus molle*** (Smith, M.J., pers. comm.); ***Carya glabra*** (Tuttle, J.P., pers. comm.); ***Rhus typhina*** (Williams, B.D., pers. comm.); ***Rhus laurina*** (Wolfe, K.L., pers. comm.).

splendens splendens (Druce) - ***Schinus molle*** (Vasquez, 1944); ***Gossypium thurberi, Juglans*** (Gardiner, 1982); ***Carya illinoensis*** (Albaugh, 1986).

vogleri (Weyenbergh) - ***Lithrea brasiliensis*** (Dias, 1978); ***Schinus dependens, Phrygilanthus cuneifolius, P. eugenioides*** (Gardiner, 1982); ***Loranthus eugenioides, L. acutifolius, Eucalyptus, Lithrea molleoides*** (Lemaire, 1988).

CITIOICA Travassos & Noronha

anthonilis (Herrich-Schäffer) - ***Salix capraea, Robinia pseudoacacia*** (Lemaire, 1988).

COLORADIA Blake

doris Barnes - ***Pinus ponderosa*** (Ferguson, 1971-72); ***Pinus strobus*** (Tuttle, J.P., pers. comm.).

luski Barnes & Benjamin - ***Pinus strobus*** (Tuttle, J.P., pers. comm.).

pandora Blake ssp. - ***Pinus ponderosa*** (Aldrich, 1912); ***Pinus echinata, Populus tremuloides*** (Tietz, 1972).

pandora davisi Barnes & Benjamin - ?***Pinus edulis*** (Ferguson, 1971-72); ***Pinus ponderosa, P. sylvestris*** (Tuttle, J.P., pers. comm.).

pandora lindseyi Barnes & Benjamin - ***Pinus jeffreyi*** (Aldrich, 1921); ***Pinus murrayana*** (=***P. contorta*** var. ***murrayana***), ***P. ponderosa*** (Patterson, 1929); ***Pinus sylvestris*** (Tuttle, J.P., pers. comm.); ***Pinus thunbergiana*** (Wolfe, K.L., pers. comm.).

pandora pandora Blake - ***Pinus ponderosa*** (Packard, 1914); ***Pinus contorta*** (Wyrant, 1941); ***Pinus contorta*** var. ***latifolia*** (Ferguson, 1971-72); ***Pinus strobus, P. sylvestris*** (Tuttle, J.P., pers. comm.).

velda J. W. Johnson & Walter - ***Pinus monophylla*** (Tuskes, 1984); ***Pinus edulis, P. sylvestris*** (Smith, M.J., pers. comm.); ***Pinus strobus*** (Tuttle, J.P., pers. comm.); ***Pinus jeffreyi, P. thunbergiana*** (Wolfe, K.L., pers. comm.).

COPAXA Walker

andensis Lemaire - ***Persea*** (Lemaire, C., pers. comm.); ***Persea americana*** (Wolfe, K.L., pers. comm.).

canella Walker - ***Phoebe porphyria*** (Bouvier, 1936).

cydippe (Druce) - ***Pinus nigra, P. radiata*** (Wolfe, K.L., pers. comm.).
decrescens Walker - ***Salix*** (Herbin, D., pers. comm.); ***Persea americana*** (Wolfe, K.L., pers. comm.).
denda Druce - ***Persea*** (Lemaire, C., pers. comm.); ***Persea americana*** (Wolfe, K.L., pers. comm.).
escalantei Lemaire - ***Persea*** (Lemaire, C., pers. comm.); ***Persea americana*** (Wolfe, K.L., pers. comm.).
joinvillea Schaus - ***Persea*** (Lemaire, C., pers. comm.).
lavendera (Westwood) - ***Quercus*** (Crotch, 1956); ***Salix babylonica, Persea*** (Lemaire, C., pers. comm.); ***Rhus trilobata*** (Stone, S.E., pers. obs.); ***Persea americana, Rhus laurina, Cinnamomum camphora*** (Wolfe, K.L., pers. comm.).
mannana Dyar - ***Quercus gambelii, Q. robur*** (Stone, S.E., pers. obs.); ***Quercus agrifolia, Q. ilex, Q. kelloggii*** (Wolfe, K.L., pers. comm.).
mazaorum Lemaire - ***Persea americana, Prunus*** (Wolfe, K.L., pers. comm.).
medea (Maassen) - ***Pinus radiata*** (Gara & Onore, 1989).
moinieri Lemaire - ***Ocotea veraguensis*** (Janzen, 1984a).
muellerana (Dyar) - ***Quercus*** (Smith, M.J., pers. comm.).
multifenestrata multifenestrata (Herrich-Schäffer) - ***Persea americana*** (Gara & Onore, 1989).
multifenestrata (Herrich-Schäffer) ssp. - ***Salix capraea*** (Naumann, S., pers. comm.); ***Persea americana*** (Wolfe, K.L., pers. comm.).
rufinans Schaus - ***Rhus laurina, Prunus, Quercus, Persea americana*** (Wolfe, K.L., pers. comm.).
semioculata orientalis Lemaire - ***Brachyotum ledifolium*** (Onore, G., pers. comm., via Rawlins, J. E.).
syntheratoides W. Rothschild - ***Crataegus*** (Lemaire, C., pers. comm.).

COPIOPTERYX Duncan
semiramis (Cramer) - ***Manilkara chicle*** (Janzen, 1984a).

COSCINOCERA Butler
hercules (Miskin) - ***Homalanthus populifolia, Panax elegans, Dysoxylum*** (Crotch, 1956); ***Timonius, Polyscias, Glochidion*** (Peigler, 1989); ***Ligustrum, Syringa vulgaris, Prunus serotina*** (Sblendorio, D.C., pers. comm.).

CRICULA Walker
andrei Jordan - ***Rhododendron ponticum*** (Packard, 1914); ***Prunus, Quercus, Pyrus malus, P. communis, Crataegus, Ligustrum, Salix*** (Crotch, 1956); "fruit trees" (Zhu & Wang, 1983).
trifenestrata (Helfer) - ***Canarium commune, Mangifera indica, Persea americana, Cinchona succirubra, Eugenia aquea, E. malaccense, Psidium guajava*** (Dupont & Scheepmaker, 1936); ***Anacardium occidentale, Careya arborea, Pyrus communis, P. malus, Machilus odoratissima, Salix babylonica, Schleichera oleosa*** (Arora &

Gupta, 1979); ***Rhododendron*, *Eucalyptus gunnii*, *Quercus*** (Gardiner, 1982); ***Canarium album*** (Zhu & Wang, 1983); ***Spondias*, *Bischofia*, *Cinnamomum*, *Artocarpus*, *Ziziphus*, *Prunus*** (Holloway, 1987).

DECACHORDA Aurivillius

aspersa Bouvier - *Chloris virgata* (Pinhey, 1972).

fulvia (Druce) - "Broad-leafed lawn grasses" (Pinhey, 1972).

rosea Aurivillius - "Grasses", *Hyparrhenia* (Pinhey, 1972).

DIRPHIA Hübner

avia (Stoll) - *Fagus sylvatica*, *Crataegus oxyacantha*, *Quercus ilex*, *Malus*, *Prunus*, *Salix* (Gardiner, 1974); *Hymenaea courbaril*, *Cedrela odorata* (Janzen, 1984a).

baroma Schaus - *Fagus sylvatica*, *Quercus ilex* (Gardiner, 1974).

curitiba Draudt - *Quercus*, *Prunus persica*, *Fagus*, JUGLANDACEAE (Gardiner, 1982).

jorgenseni Schaus - See ***Dirphiopsis ursina*** Walker (Lemaire, C., pers. comm.).

panamensis Schaus - *Salix* (Herbin, D., pers. comm.); *Robinia pseudoacacia*, *Schinus terebinthifolius* (Wolfe, K.L., pers. comm.).

tarquinia (Cramer) - "Locos tree" (Packard, 1914).

DIRPHIOPSIS Bouvier

delta (Foetterlé) (=***lombardi*** Bouvier) (Lemaire, C., pers. comm.) - ?*Prunus* (Gardiner, 1982).

ursina Walker - *Wisteria*, *Schinus gracilipes* (Gardiner, 1982).

DRYOCAMPA Harris

rubicunda (Fabricius) ssp. - *Acer saccharinum* (Holland, 1903); *Acer negundo*, *A. rubrum*, *Fagus grandifolia*, *Juglans cinerea*, *Quercus coccinea*, *Q. ilicifolia*, ?*Q. velutina* (Tietz, 1972).

rubicunda rubicunda (Fabricius) - *Acer rubrum* (Packard, 1893); *Acer spicatum*, *A. saccharinum*, *A. saccharum* (McGugan *et al.*, 1958); *Quercus laevis* (Ferguson, 1971-72); *Platanus* (Gardiner, 1982); *Acer monspessulanum* (Lemaire, 1988).

rubicunda alba Grote - *Acer saccharinum* (Ferguson, 1971-72).

DYSDAEMONIA Hübner

boreas (Cramer) - "Paineira" (Packard, 1914); *Bombax ceiba* (Crotch, 1956); *Ceiba pentandra*, MALVACEAE, TILIACEAE (Janzen, 1984a); *Chorisia* (Wolfe, K.L., pers. comm.).

fosteri W. Rothschild - *Chorisia speciosa*, ?MALVACEAE, ?TILIACEAE (Crotch, 1956); *Samanea* (=*Albizia*) (Moucha, 1966).

EACLES Hübner

canaima Feige - CORNACEAE (Lemaire, 1988).

ducalis Walker - MALVACEAE, MELASTOMATACEAE, *Ligustrum*, *Juglans*, *Fagus*, *Quercus* (Lemaire, 1988).

imperialis (Drury) ssp. - *Platanus occidentalis*, *Quercus*, *Liquidambar styraciflua*, *Pinus* (Morris, 1862); *Prunus virginiana* (Packard, 1893); *Acer*, *Cedrus*, *Prunus* (Forbes, 1923); *Ligustrum*, *Ulmus*, *Fagus*, *Castanea* (Crotch, 1956); *Diospyros* (Baker, 1972); *Abies grandis*, *Acer negundo*, *A. pseudoplatanus*, *A. rubrum*, *A. saccharinum*, *A. saccharum*, *Aesculus hippocastanum*, *Alnus*, *A. serrulata*, *Betula alba*, *B. alba* var. *laciniata*, *B. papyrifera*, *B. populifolia*, *Carpinus caroliniana*, *Carya*, *Castanea dentata*, *C. pumila*, *C. sativa*, *Chamaecyparis thyoides*, *Cornus*, *Crataegus*, *Fagus grandifolia*, *F. sylvatica*, *Gleditsia triacanthos*, *Gossypium herbaceum*, *Juglans cinerea*, *Juniperus communis*, *J. virginiana*, *Koelreuteria paniculata*, *Larix laricina*, *Lindera benzoin*, *Myrica cerifera*, *Ostrya virginiana*, *Picea*, *Pinus banksiana*, *P. echinata*, *P. griffithii*, *P. strobus*, *Platanus orientalis*, *Prunus serotina*, *Quercus coccinea*, *Q. laevis*, *Q. macrocarpa*, *Q. palustris*, *Q. rubra*, *Rhus glabra*, *Rosa*, *Salix*, *Sambucus canadensis*, *Sassafras albidum* var. *molle*, *Taxodium distichum*, *Tilia*, *T. americana*, *T. europaea*, *T. heterophylla*, *Tsuga canadensis*, *Ulmus alata*, *U. americana*, *U. procera*, *U. rubra*, *U. thomasii*, *Viburnum dentatum* (Tietz, 1972); *Cochlospermum vitifolium*, *Bursera tomentosa* (Janzen, 1982); *Platanus*, *Rhus*, *Ulmus* (Albaugh, 1986); *Pinus taeda* (Furr, M., pers. comm.); *Pinus serotina* (Williams, B.D., pers. comm.); *Robinia*, *Rhus laurina* (Wolfe, K.L., pers. comm.).

imperialis anchicayensis Lemaire - *Quercus robur* (Naumann, S., pers. comm.).

imperialis cacicus (Boisduval) - *Mangifera indica*, *Erythrina crista-galli* (Packard, 1905); *Quercus ilex*, *Pistacia lentiscus* (Lemaire, 1988).

imperialis decoris W. Rothschild - *Cochlospermum vitifolium*, *Bursera tomentosa*, *Astronium graveolens*, *Cedrela odorata* (Janzen, 1984a); *Quercus* (Purdum, D.A., pers. comm.).

imperialis imperialis (Drury) - *Carya*, *Juniperus*, *Pinus rigida* (Eliot & Soule, 1902); *Quercus*, *Platanus*, *Tilia*, *Acer saccharinum*, *Gleditsia triacanthos*, *Prunus virginiana*, *Rhus*, *Sassafras albidum*, *Ulmus*, *Fagus*, *Alnus*, *Pinus*, *Picea*, *Betula*, *Carpinus*, *Cupressus*, *Juglans regia*, *Liquidambar styraciflua*, *Tsuga* (Ferguson, 1971-72).

imperialis opaca (Burmeister) - *Erythrina crista-galli*, *Salix babylonica* (Lemaire, 1988); *Quercus robur* (Naumann, S., pers. comm.); *Prunus serotina* (Purdum, D.A., pers. comm.); *Ligustrum* (Stone, S.E., pers. obs.).

imperialis oslari [2] W. Rothschild - *Juglans* (Gardiner, 1982); *Quercus oblongifolia*, *Q. emoryi*, *Schinus molle*, *Liquidambar styraciflua* (Tuskes, 1986b); *Robinia neomexicana* (Sblendorio, D.C., pers. comm.); *Quercus chihuahuaensis* (Smith, M.J., pers. comm.); *Quercus robur*, *Q. gambelii* (Stone, S.E., pers. obs.);

---

[2] Lemaire (1988: 34) lists several conifers as foodplants for *E. imperialis oslari*. However, Gage (1976: 175-176) records high larval mortality on conifers.

Quercus rubra (Williams, B.D., pers. comm.); *Rhus laurina* (Wolfe, K.L., pers. comm.).

imperialis pini Michener - *Pinus banksiana, P. strobus* (Michener, 1950); *Pinus resinosa* (McGugan *et al.*, 1952); *Pinus sylvestris, Picea glauca* (Ferguson, 1971-72); *Sapindus saponaria, Quercus palustris* (Tuttle, J.P., pers. comm.).

imperialis tucumana W. Rothschild - *Ligustrum* (Crotch, 1956); *Schinus dependens, Celtis* (Lemaire, 1988).

masoni Schaus ssp. - *Rhus laurina* (Wolfe, K.L., pers. comm.).

ormondei Schaus ssp. - *Quercus, Rhus laurina* (Wolfe, K.L., pers. comm.).

penelope (Cramer) - *Psidium pomiferum*, MELASTOMATACEAE (Packard, 1905); *Quercus ilex, Fagus sylvatica* (Gardiner, 1966); *Laburnum, Ligustrum, Juglans regia* (Gardiner, 1982); *Salix* (Lemaire, 1988).

EOCHROA Felder

trimeni Felder - *Melianthus minor* (Pinhey, 1972); *Melianthus pectinatus* (Oberprieler, 1987); *Melianthus dregianus* (Wolfe, K.L., pers. comm.).

EPIPHORA Wallengren

albida (Druce) - *Paullinia, Fagara macrophylla* (Gardiner, 1982).

atbarina Butler - *Ziziphus* (Gardiner, 1982).

bauhiniae (Guérin-Méneville) - *Ziziphus mucronata, Z. spinachristii, Z. jujuba, Z. mauritiana* (Pinhey, 1972).

boolana Strand - *Discoglypremna caloneura* (Rougeot, 1962).

lugardi Kirby - *Discoglypremna caloneura* (Gardiner, 1982).

mythimnia (Westwood) - "Psekamafoura", *Croton sylvaticus, Helinus ovatus, Ziziphus mucronata* (Platt, 1921); *Croton gratissimus* (Oberprieler, 1987); *Rhamnus frangula* (Naumann, S., pers. comm.).

ploetzi (Plötz) - *Fagara macrophylla* (Gardiner, 1982).

rectifascia Rothschild - *Ziziphus mauritiana, Maesopsis emini* (Crotch, 1956); *Ziziphus mauretanicus* (Rougeot, 1962); *Croton* (Gardiner, 1982).

vacuna (Westwood) - *Maesopsis emini* (Pinhey, 1972); *Fagara heitzii, Croton oligondrum* (Gardiner, 1982).

ERIOGYNA Jordan

pyretorum Westwood - *Liquidambar formosana, Cinnamomum camphora* (Crotch, 1956); *Crataegus* (Gardiner, 1982); *Eriobotrya japonica, Pterocarya stenoptera, Rosa multifolia, Psidium guajava, Pyrus pyrifolia, Ulmus pumila* (Zhu & Wang, 1983).

EUBERGIA Bouvier

caisa (Berg) (=***boetifica*** Druce) (Lemaire, C., pers. comm.) - *Mimosa laurentzii* (Gardiner, 1982).

EUDYARIA Butler

venata Butler - *Eryngium agavifolium, Poterium, Sanguisorba, Conopodium denudatum, Pimpinella saxifraga*, UMBELLIFERAE (=APIACEAE) (Crotch, 1956).

EUPACKARDIA Cockerell

calleta (Westwood) - *Prunus serotina* (Packard, 1914); *Ligustrum, Betula* (Crotch, 1956); *Leucophyllum frutescens, Fraxinus, Syringa vulgaris, Salix, Schinus molle* (Collins & Weast, 1961); *Fouquieria splendens* (Ferguson, 1971-72); *Pyrus malus* (Gardiner, 1982); *Forestiera angustifolia* (Peigler, R.S., pers. comm.); *Sapium biloculare* (Smith, M.J., pers. comm.); *Salix babylonica* (Tuttle, J.P., pers. comm.); *Rhus laurina* (Wolfe, K.L., pers. comm.).

EUSTERA Duncan

argiphontes Maassen & Kirby - *Albizia* (Gardiner, 1982).

brachyura Drury - *Dialium guineense* (Packard, 1914).

GAMELIA Hübner

abasia (Stoll) - *Psidium guajava* (Gardiner, 1982).

GONIMBRASIA Butler

belina (Westwood) - *Carissa grandiflora, Ficus, Rhus longifolia, Sclerocarya caffra, Trema bracteolata* (Platt, 1921); *Terminalia schweinfurthi, T. velutina* (Rougeot, 1962); *Colophospermum mopane, Diospyros, Rhus lancea* (Pinhey, 1972); *Copaifera* (Gardiner, 1982); *Pinus radiata, Sclerocarya birrea, Schotia brachypetala, Burkea africana, Trema orientalis, Platanus* (Oberprieler, 1986).

hecate (Rougeot) - *Combretum, Berlinia* (Rougeot, 1962); *Terminalia, Annona, Bauhinia* (Gardiner, 1982).

osiris annulata (Druce) - *Terminalia schweinfurthi* (Bouvier, 1936).

tyrrhea (Cramer) ssp. - *Acacia mollissima* (Platt, 1921); *Acacia karroo* (Dickerson, 1945); *Pyrus malus, Crataegus* (Crotch, 1956); *Salix* (Pinhey, 1972); *Diospyros lycioides* (Oberprieler, 1987); *Quercus robur* (Naumann, S., pers. comm.); *Prunus virginiana* (Stone, S.E., pers. obs.); *Rhus laurina* (Wolfe, K.L., pers. comm.).

zambesina (Walker) - *Nerium oleander* (Platt, 1921); *Mangifera, Poinciana* (=*Delonix*), *Diospyros mespiliformis* (Pinhey, 1972); *Fagus* (Gardiner, 1982); *Rhus* (Wolfe, K.L., pers. comm.).

GOODIA Holland

kuntzei (Dewitz) - *Brachystegia randii* (Platt, 1921); *Berlinia paniculata, Brachystegia venosa* (Crotch, 1956); *Acacia sieberiana, Brachystegia spiciformis, Julbernardia paniculata* (Pinhey, 1972).

lunata Holland - *Aframomum* (Rougeot, 1962); *Ficus, Amomum* (Gardiner, 1982).

nubilata Holland - ZINGIBERACEAE, AMPELIDACEAE (=VITACEAE), STERCULIACEAE (Gardiner, 1982).

sentosa Jordan - *Aframomum* (Rougeot, 1962); *Amomum* (Gardiner, 1982).

GRAELLSIA Grote

isabellae (Graëlls) ssp. - *Pinus maritima* (Packard, 1914); *Pinus*, esp. *P. sylvestris* (Crotch, 1956); *Pinus laricio* (Rougeot & Viette, 1983).

GYNANISA Walker

ata Strand - ?*Khaya nyasica* (Pinhey, 1968); *Brachystegia venosa, Acacia, Julbernardia* (Pinhey, 1972).

maja (Klug) ssp. - *Acacia mollissima* (Platt, 1921); *Acacia karroo* (Taylor, 1947); *Berlinia paniculata, Brachystegia venosa, Laburnum* (Crotch, 1956); *Elephantorrhiza burchelli* (Rougeot, 1962); *Prunus persica* (Pinhey, 1972); *Cassia, Berlinia, Robinia pseudoacacia*, "fruit trees" (Gardiner, 1982); *Colophospermum, Peltophorum, Pterocarpus, Schotia, Julbernardia* (Oberprieler, 1987); *Quercus robur, Q. turneri* (Naumann, S., pers. comm.); *Quercus gambelii* (Stone, S.E., pers. obs.).

nigra Bouvier - *Colophospermum mopane* (Pinhey, 1968); *Acacia* (Pinhey, 1972).

HELICONISA Walker

pagenstecheri (Geyer) - "Grasses" (Gardiner, 1982).

HEMILEUCA Walker

artemis Packard - *Populus fremontii* (Packard, 1893); *Salix longifolia* (Packard, 1914).

burnsi J. H. Watson - *Eriogonum fasciculatum, Parosela (=Dalea) californica, Tetradymia spinosa* (Comstock & Dammers, 1937); *Tetradymia glabrata, Dalea fremontii* (Ferguson, 1971-72); *Tetradymia axillaris* (McFarland, 1975); *Prunus fasciculata, Eriogonum fasciculatum* var. *polifolium* (later instars) (Tuskes, 1984); *Acacia greggi* (Stone & Smith, 1990); *Gleditsia triacanthos* (Curtis, W.M., pers. comm.); *Prunus virginiana* (Stone, S.E., pers. obs.).

chinatiensis (Tinkham) - *Mimosa lindheimeri, M. borealis, Ephedra nevadensis* var. *aspera, Rhus aromatica* var. *flabelliformis, Forestiera angustifolia* (Ferguson, 1971-72); *Condalia, Acacia* (Stone & Smith, 1990); *Fremontia californica, Rhus microphylla* (Smith, M.J., pers. comm.); *Cercocarpus* (Wolfe, K.L., pers. comm.).

conwayae Peigler - *Condalia ericoides, Rhus microphylla* (Peigler, 1985).

diana Packard - *Quercus oblongifolia* (Kendall & Peigler, 1981); *Quercus emoryi* (Tuskes, 1986a); *Quercus turbinella* (Smith, M.J., pers. comm.); *Quercus gambelii, Q. robur* (Stone, S.E., pers. obs.); *Quercus lobata* (Tuskes, P.M., pers. comm.); *Quercus palustris* (Tuttle, J.P., pers. comm.).

eglanterina (Boisduval) ssp. - *Prunus, Ceanothus, Arctostaphylos, Rosa, Rhamnus purshiana* (Packard, 1914); *Ceanothus thyrsiflorus* (Tilden, 1945); *Amelanchier* (Evans, 1958); *Salix, Betula, Populus tremuloides, Acer circinatum, Holodiscus discolor* (McGugan *et al.*, 1958); *Ceanothus spinosa, Rhamnus californicus, Prunus* *virginiana, Purshia tridentata* (Ferguson, 1971-72); *Arctostaphylos glauca, A. uva-ursi, Prunus cerasus, P. domestica, P. domestica* var. *gelatensis, P. serotina, Rhamnus alnifolia, Rubus allegheniensis, Vitis californica, V. vulpina*, ROSACEAE (Tietz, 1972); *Cercocarpus betuloides* (McFarland, 1974); *Symphoricarpos* (Tuskes, 1976); *Pyrus malus, Acer, Symphoricarpos albus, S. mollis, Spiraea menziesii* (Gardiner, 1982); *Pyracantha, Rubus, ?Quercus, Prunus emarginata, Ceanothus velutinus, Symphoricarpos vaccinoides* (Tuskes, 1984); *Prunus avium* (Naumann, S., pers. comm.); *Adenostoma fasciculatum* (Smith, M.J., pers. comm.); *Cercocarpus montanus* (Stone, S.E., pers. obs.); *Potentilla* (Warren, A.D., pers. comm.); *Salix discolor* (Williams, B.D., pers. comm.).

electra W. G. Wright ssp. - *Eriogonum fasciculatum* (Comstock & Dammers, 1939); *Eriogonum fasciculatum* var. *polifolium* (Smith, 1974); *Bursera microphylla, Idria columnaris, Tamarix gallica, Simmondsia chinensis* (Stone & Smith, 1990).

griffini Tuskes - *Coleogyne ramosissima, Cercocarpus betuloides* (Tuskes, 1978); *Amelanchier* (Smith, 1988); *Ephedra* (Stone & Smith, 1990).

grotei Grote & Robinson - *Quercus virginiana* var. *fusiformis, Q. virginiana* (Ferguson, 1971-72); *Quercus havardii* X *Q. stellata, Q. marilandica, Q. texana* (Kendall & Peigler, 1981); *Quercus gambelii* (Stone, S.E., pers. obs.); *Quercus stellata* (Tuttle, J.P., pers. comm.).

hera hera (Harris) - *Artemisia* (Packard, 1914); *Artemisia tridentata* (McFarland, 1974); *Lupinus, Eriogonum* (Tuskes, 1976); *Artemisia cana* (Collins, M.M., pers. comm.).

hera marcata (Neumoegen) - *Artemisia tridentata* (Stone *et al.*, 1988).

hualapai (Neumoegen) - *Bouteloua curtipendula* (Ferguson, 1971-72); *Cynodon* (Smith, M.J., pers. comm.).

juno Packard - *Prosopis juliflora* (Comstock & Dammers, 1939); *Prosopis glandulosa* (Tietz, 1972); *Gleditsia triacanthos* (Tuttle, J.P., pers. comm.).

lex (Druce) - "Leguminous plants", *Robinia pseudoacacia, Fagus* (Gardiner 1982); *Quercus arizonica, Eriogonum wrightii* (Hansen, K.C., pers. comm.).

lucina Henry Edwards - *Spiraea latifolia* (Reiff, 1910); *Betula populifolia, Prunus serotina, Quercus, Spiraea salicifolia* (Tietz, 1972); *?Salix* (Peigler & Stone, 1989); *Spiraea tomentosa* (Tuttle, J.P., pers. comm.).

magnifica (Rotger) - *Artemisia filifolia* (Townsend, 1893 [as *Hemileuca maia*]); *Artemisia tridentata* (Stone *et al.*, 1988); *Eriogonum effusum* (Beutler, L., pers. comm.).

maia (Drury) - *Quercus rubra* var. *ambigua* (Smith, 1797); *Prunus, Salix, Corylus* (Holland, 1903); *Quercus velutina* (Ely, 1954); *Quercus ilicifolia, Q. shumardii, Q. virginiana* (Ferguson, 1971-72); *Populus tremuloides, Prunus cerasus, P. serotina, Quercus laevis, Q. stellata* (Tietz, 1972); *Salix babylonica* (Leeuw, 1974); *Salix hindsiana* (Smith, 1974); *Quercus prinoides* (Cryan & Dirig, 1977); *Populus fremontii* (Stone, S.E., pers. obs.); *Salix interior, Betula pumila* (Tuttle, J.P., pers. comm.).

mania (Druce) - "Pasture grasses", *Zea mays* (Crotch, 1956).

neumoegeni Henry Edwards - ***Eriogonum fasciculatum*** (Bauer, 1948); ***Prunus fasciculata, Rhus trilobata*** (Tuskes, 1976); ***Fallugia paradoxa*** (Stone & Smith, 1990); ***Prunus virginiana*** (Stone, S.E., pers. obs.).

nevadensis Stretch - ***Populus fremontii, Salix longifolia*** (Packard, 1914 [as ***Hemileuca maia***]); ***Populus tremuloides*** (McGugan *et al.*, 1958); ***Prunus*** (Ferguson, 1971-72); ***Salix interior, S. lasiandra, S. lasiolepis*** (Tietz, 1972); ***Corylus*** (Villiard, 1969); ***Salix babylonica*** (Tuttle, J.P., pers. comm.); ***Salix discolor*** (Williams, B.D., pers. comm.); ***Quercus agrifolia*** (Wolfe, K.L., pers. comm.).

nuttalli nuttalli (Strecker) - ***Purshia tridentata, Symphoricarpos*** (Ferguson, 1971-72); *Cercocarpus betuloides* (McFarland, 1974); *Symphoricarpos oreophilus* (Richards, J.L., pers. comm.); *Pyracantha* (Smith, M.J., pers. comm.); *Symphoricarpos albus* (Stone, S.E., pers. obs.).

nuttalli uniformis (Cockerell) - *Symphoricarpos* (Ferguson, 1971-72); *Purshia tridentata, Symphoricarpos vaccinoides* (Tuskes, 1984).

oliviae Cockerell - ***Bouteloua hirsuta, B. oligostachya, B. curtipendula, Buchloe dactyloides, Hilaria cenchroides, Andropogon hallii, Lycurus phleoides, Distichlis spicata, Sitanion jubatum, Hordeum caespitosum, Bromus polyanthus*** (Ainslie, 1910); "a grass-feeder", ***Muhlenbergia*** ?*texana* (Packard, 1914); *Agropyron smithii, Bouteloua gracilis, Sporobolus airoides* (Tietz, 1972); *Sorghum* (Peigler, R.S., pers. comm.).

peigleri Lemaire - *Quercus havardii* (Lemaire, 1979a); *Quercus fusiformis, Q. texana* (Peigler & Stone, 1989); *Quercus robur* (Stone, S.E., pers. obs).

rubridorsa Felder - "Leguminous plants", *Mimosa dealbata, Robinia pseudoacacia, Salix, Quercus* (Gardiner, 1982); *Acacia* (Wolfe, K.L., pers. comm.).

slosseri Peigler & Stone - *Quercus havardii, Q. fusiformis* (Peigler & Stone, 1989); *Quercus macrocarpa* (Peigler, R.S, pers. comm.); *Quercus robur* (Stone, S.E., pers. obs.).

tricolor (Packard) - *Cercidium* (Bauer, 1948); *Cercidium microphyllum, Idria columnaris, Prosopis* (Stone & Smith, 1990); *Gleditsia triacanthos, Quercus* (Smith, M.J., pers. comm.); *Robinia neomexicana, Quercus rubra* (Stone, S.E., pers. obs.); *Mimosa* (Wolfe, K.L., pers. comm.).

HENIOCHA Hübner

apollonia (Cramer) - *Acacia mollissima* (Platt, 1921).

dyops (Maassen & Weymer) - *Acacia mollissima* (Platt, 1921).

marnois Rogenhofer - *Acacia mollissima* (Pinhey, 1972).

HOLOCERINA Pinhey

agomensis (Karsch) - *Uapaca, Brachystegia, Bauhinia reticulata* (Crotch, 1956); *Monotes glaberi, Protea, Uapaca kirkiana, U. nitida* (Pinhey, 1972); *Laburnum* (Gardiner, 1982).

angulata (Aurivillius) - *Bauhinia reticulata* (Crotch, 1956); *Spathodea campanulata* (Rougeot, 1962); *Psorospermum, Piliostigma* (=*Bauhinia*) (Pinhey, 1972).

rhodesiensis (Janse) - *Cussonia spicata* (Platt, 1921).
smilax (Westwood) - *Jasminum pubigerum, Quercus* (Packard, 1914); *Rhus laevigata, R. longifolia, R. villosa, Psidium, Ekebergia meyeri, Quercus pedunculata* (Platt, 1921); *Ligustrum, Psidium guajava* (Pinhey, 1972); *Pinus patula* (Van den Berg & Van den Berg, 1973); *Piliostigma* (=*Bauninia*) (Pinhey, 1975); *Rhus trilobata* (Gardiner, 1982); *Pyrus malus, Syringa vulgaris, Rhus glabra* (Peigler, R.S., pers. comm.); *Quercus gambelii* (Stone, S.E., pers. obs.).

HOMOEOPTERYX Jordan
RUBIACEAE (Crotch, 1956).

HYALOPHORA Duncan
cecropia (Linnaeus) - *Pyrus malus, Prunus, Ribes, Berberis* (Morris, 1862); *Tilia americana, T. europaea, Acer spicatum, A. saccharinum, A. dasycarpum, A. rubrum, Negundo aceroides, Prunus americana, P. domestica, P. pennsylvanica, P. serotina, P. vulgaris, Spiraea opulifolia, S. salicifolia, S. tomentosa, Crataegus coccinea, C. tomentosa, C. crus-galli, Pyrus communis, P. coronaria, P. americana, Amelanchier canadensis, Ribes floridum, R. floridum* var. *officinale, R. rubrum* var. *officinale, Sambucus canadensis, S. pubens, Ulmus americana, U. racemosa, U. fulva, Quercus alba, Fagus ferruginea, Corylus americana, C. rostrata, Carpinus americana, Betula alba, B. excelsa, B. lenta, Alnus incana, A. serrulata, Salix humilis, S. lucida, S. viminalis, S. alba, Populus tremuloides, P. balsamifera* (Brodie, 1882)[3]; "a predilection for the ROSACEAE" (Holland, 1903); *Syringa vulgaris, Alnus glutinosa, Ribes rubrum, Carya, Gleditsia triacanthos* (Saunders, 1909); *Fraxinus, Populus, Rosa, Liriodendron, Rubus, Ceanothus, Vaccinium, Acer plantanoides,* "fruit trees" (Packard, 1914); *Acer negundo* (Priddle, 1966); *Myrica cerifera, M. carolinensis, Spiraea latifolia, Viburnum cassinoides, Betula populifolia* (Ferguson, 1971-72); *Juglans regia, Salix discolor* (Baker, 1972); *Carya illinoensis* (Brown, 1972); *Aronia arbutifolia, Berberis vulgaris, Betula alleghaniensis, B. papyrifera, Cephalanthus occidentalis, Corylus cornuta, Crataegus calpodendron, C. oxyacantha, C. pedicellata, Gaylussacia frondosa, Humulus lupulus, Malus coronaria, M. pumila, Ostrya virginiana, Paeonia officinalis, Parthenocissus quinquefolia, Philadelphus inodorus, Physocarpus opulifolius, Prunus cerasus, P. ilicifolia, P. maritima, P. virginiana, Ribes americanum, R. grossularia, R. nigrum, R. sativum, Rosa blanda, Rubus allegheniensis, R. idaeus* var. *strigosus, R. occidentalis, Sambucus racemosa* var. *pubens, Sassafras albidum* var. *molle, Sorbus americana, Spiraea betulifolia* var. *corymbosa, Symphoricarpos albus, Ulmus rubra, U. thomasii, Wisteria sinensis* (Tietz, 1972); *Cornus stolonifera* (Waldbauer & Sternburg, 1973); *Larix*

---

[3] Packard (1914: 212) incorrectly gives this citation as Brodie (Papilio, Feb., 1882, v. 1, p. 32-33). The correct citation is in References Consulted here.

(Scarbrough *et al.*, 1974); ***Rhus glabra, Salix interior, Toxylon*** (=*Maclura*) ***pomiferum*** (Sternburg & Waldbauer, 1981); ***Cephalanthus*** (Albaugh, 1986); ***Prunus avium*** (Naumann, S., pers. comm.); ***Morus alba, Rhus typhina*** (Sblendorio, D.C., pers. comm.); ***Ribes alpinum*** (Stone, S.E., pers. obs.); ***Rhus laurina*** (Wolfe, K.L., pers. comm.).

columbia (S. I. Smith) ssp. - ***Abies balsamea, Myrica pensylvanica, Picea glauca, P. marina, Prunus cerasus, P. pensylvanica, P. virginiana, Salix alba, Spiraea tomentosa, Tsuga canadensis, Ulmus*** (Tietz, 1972); ***Prunus serotina*** (Sblendorio, D.C., pers. comm.).

columbia columbia (S. I. Smith) - ?*Nemopanthus canadensis* (=*Nemopanthus mucronata*), ?*Rhodora canadensis* (=***Rhododendron*** *canadense*), ?*Kalmia* ***angustifolia***, ?***Acer, Larix laricina*** (Smith, 1865)[4]; ***Prunus virginiana*** (Collins & Weast, 1961); ***Salix babylonica*** (Tuttle, 1985); ***Prunus domestica, Pseudotsuga menziesii*** (Stone, S.E., pers. obs.).

columbia gloveri (Strecker) - ***Prunus serotina, P. virginiana*** (Eliot & Soule, 1902); ***Salix*** (Comstock, 1931); ***Larix decidua*** (Crotch, 1956); ***Acer, Alnus, Elaeagnus angustifolia, Ribes, Salix exigua*** (Collins & Weast, 1961); ***Ligustrum, Syringa vulgaris*** (Villiard, 1969); ***Shepherdia*** (Ferguson, 1971-72); ***Acer negundo, Larix laricina, Ribes lacustre, R. sativum, Ulmus*** (Tietz, 1972); ***Pseudotsuga*** (Collins, 1973); ***Purshia, Rosa*** (Tuskes, 1976); ***Arctostaphylos patula, Ceanothus cordulatus, C. velutinus, Prunus demissa, P. emarginata, Purshia tridentata, Shepherdia argentea*** (Collins, 1984); ***Prunus domestica, Betula pendula*** (Naumann, S., pers. comm.); ***Arctostaphylos*** (Smith, M.J., pers. comm.); ***Ribes alpinum*** (Stone, S.E., pers. obs.); ***Salix babylonica, Rhus trilobata*** (Tuttle, J.P., pers. comm.); ***Salix discolor*** (Williams, B.D., pers. comm.).

columbia nokomis (Brodie) - ***Salix alba*** (Skinner, 1908); ***Elaeagnus argentea*** (Packard, 1914).

euryalus euryalus (Boisduval) - ***Ceanothus thyrsiflorus*** (Holland, 1903); ***Prunus serotina*** (Packard, 1914); ***Amelanchier alnifolia, Ceanothus sanguineus, Acer glabrum, Ribes sanguineum*** (Jones, 1951); ***Salix*** (Crotch, 1956); ***Alnus rubra, Rosa, Shepherdia canadensis*** (McGugan *et al.*, 1958); ***Arctostaphylos, Betula occidentalis, Pyrus malus, Schinus molle, Cephalanthus, Rhamnus californicus, Ribes*** (Collins & Weast, 1961); ***Malus pumila, Syringa vulgaris*** (Villiard, 1969); ***Amelanchier florida, Arctostaphylos glauca, Ceanothus integerrimus, Larix laricina, Photinia arbutifolia, Picea, Prunus virginiana, Pseudotsuga menziesii, Rhamnus californicus*** var. ***tomentella, R. crocea, R. purshiana*** (Tietz, 1972); ***Rhus, Arbutus, Quercus*** (Tuskes, 1976); ***Purshia tridentata*** (Furniss & Carolin, 1977); ***Arctostaphylos patula, Ceanothus cordulatus, C. velutinus, Prunus demissa, P. emarginata, Salix exigua, Shepherdia argentea*** (Collins, 1984); ***Coffea*** (Albaugh,

---

[4] Except for *Larix laricina* these foodplants are probably erroneous, as explained by Ferguson (1971-72: 253).

1986); ***Ribes alpinum*** (Stone, S.E., pers. obs.); ***Salix babylonica*** (Tuttle, J.P., pers. comm.); ***Salix discolor*** (Williams, B.D., pers. comm.); ***Rhus laurina, Liquidambar styraciflua*** (Wolfe, K.L., pers. comm.).

euryalus cedrosensis (Cockerell) - *Pachycormus discolor, Rhus laurina, R. integrifolia, Schinus molle* (Smith, M.J., pers. comm.).

HYLESIA Hübner

acuta Druce - *Schinus terebinthifolius* (Wolfe, 1988); *Bursera simaruba* (Wolfe, K.L., pers. comm.).

alinda Druce - *Persea americana* (Gardiner, 1982).

continua cressida Dyar - *Quercus* (Smith, M.J., pers. comm.).

dalina Schaus - *Casearia odorata, Malvaviscus arboreus* (Janzen, 1984a); *Casearia arguta* (Janzen, 1985).

lineata Druce - *Tabebuia rosea, Bombacopsis quinatum, Cordia alliodora, Hirtella racemosa, Muntingia calabura, Casearia arguta, C. corymbosa, C. sylvestris, Zuelania guidonia, Acacia tenuifolia, Cassia biflora, Diphysa robinioides, Enterolobium cyclocarpum, Hymenaea courbaril, Lonchocarpus minimiflorus, L. costaricensis, Lysiloma auritum, Machaerium kegelii, Mimosa pigra, Myrospermum frutescens, Pithecellobium lanceolatum, Hyptis pectinata, Malvaviscus arboreus, Banisteriopsis muricata, Byrsonima crassifolia, Stigmaphyllon ellipticum, Psidium guineense, Ouratea lucens, Gouania polygama, Calycophyllum candidissimum, Chomelia spinosa, Guettarda macrosperma, Xanthoxylum setulosum* (=*Zanthoxylum*), *Allophyllus occidentalis, Cupania guatemalensis, Paullinia cururu, Serjania schiedeana, Urvillea ulmacea, Byttneria aculeata, B. catalpaefolia, Guazuma ulmifolia, Luehea speciosa, Lantana camara, Erythroxylum havanense, Calliandra emarginata* (Janzen, 1984a); *Bauhinia ungulata, ?Persea americana, ?Dalbergia retusa, ?Gliricidia sepium, ?Inga vera, ?Pithecellobium lanceolatum, ?Triumfetta lappula* (Janzen, 1984b); *Quercus* (Smith, M.J., pers. comm.); *Acacia baileyana, Robinia pseudoacacia* (larvae are nocturnal feeders, Wolfe, K.L., pers. comm.).

nigricans Berg - "Fruit trees", *Salix* (Crotch, 1956); *Populus, Pyrus malus, Quercus* (Gardiner, 1982).

remex Dyar - *Porlieria lorentzii* (Gardiner, 1982).

wagneri Bouvier - MELASTOMATACEAE, *Tibouchina holosericea* (Gardiner, 1982).

HYPERCHIRIA Hübner

incisa (Walker) ssp. - *Quercus, Q. ilex* (Crotch, 1956); *Cassia corymbosa, Celtis spinosa, Ficus benjamina, Platanus orientalis, Wisteria, Berberis lauina, Serjania, Fagus, Laburnum* (Gardiner, 1982).

nausica (Cramer) - *Anacardium, Laburnum, Robinia pseudoacacia, Fagus* (Gardiner, 1982).

orodina (Schaus) - SAPINDACEAE, *Quamoclit setosa* (Gardiner, 1982).

IMBRASIA Hübner

anna (Maassen & Weymer) - ***Hyphaene crinita*** (Platt, 1921).

carnegiei (Janse) - ***Brachystegia globifera*** (Platt, 1921); ***Mimosa richelmanni, Brachystegia*** (Bouvier, 1936); ***Protea*** (Pinhey, 1972).

cleoris (Jordan) - ***Prunus, Crataegus, Rosa, Ligustrum*** (Crotch, 1956).

cytherea (Fabricius) ssp. - ***Acacia cyclopis, Euclea racemosa*** (Dickerson, 1944); ***Euclea polyandra*** (Dickerson, 1945); ***Cydonia, Leucospermum ellipticum, Psidium, ?Pyrus communis, Watsonia iridifolia, Myrsine melanophloeos, Rhus angustifolia, R. crenata, R. tomentosa, R. lucida, Protea mellifera, P. mundii, Acacia saligna, A. cyanophylla, Myrica conifera, Rhamnus prunoides*** (Pinhey, 1972); ***Crataegus, Pinus, Pyrus malus, Quercus, Rhamnus*** (Pinhey, 1975); ***Myrica cordifolia, Acacia mollissima, Cupressus macrocarpa, Eucalyptus globulus, Euclea daphnoides, Liriodendron tulipifera*** (Gardiner, 1982); ***Pinus sylvestris*** (Naumann, S., pers. comm.).

cytherea clarki (Geertsema) - ***Rhus dura*** (Van den Burg & Van den Burg, 1973); ***Pinus radiata*** (Van den Berg, 1974).

dione (Fabricius) - ***Ricinus communis, Mimosa, Theobroma cacao*** (Bouvier 1936); ***Jatropha curcas, Spondias lutea, Anacardium occidentale*** (Rougeot, 1962); ***Aleurites fordii, A. montana, Chlorophora, Bauhinia petersiana, Diplorhynchus mossambicensis, Eugenia, Khaya, Psidium*** (Pinhey, 1972); ***Erythroxylum havanense*** (Gardiner, 1982).

emini (Butler) - ***Protea affinis*** (Gardiner, 1982).

ertli Rebel - ***Brachystegia randii*** (Platt, 1921); ***Julbernardia paniculata, Brachystegia spiciformis, Acacia*** (Pinhey, 1972); ***Berlinia, Laburnum, Robinia pseudoacacia*** (Gardiner, 1982).

gueinzii (Staudinger) - ***Maesa alnifolia*** (Platt, 1921); ***Myrica*** (Bouvier, 1936); ***Pyrus malus, Prunus*** (Crotch, 1956).

krucki (Hering) - ***Schinus, Acocanthera, Fagus, Betula*** (Gardiner, 1982).

longicaudata (Holland) - "Polyphagous" (Gardiner, 1982).

macrops (Rebel) - ***Brachystegia venosa, Pinus*** (Pinhey, 1972).

macrothyris Rothschild - ***Brachystegia randii*** (Platt, 1921); ***Prunus, Pyrus malus*** (Crotch, 1956); ***Julbernardia, Brachystegia spiciformis*** (Pinhey, 1972).

nictitans (Fabricius) - ***Annona senegalensis, Bauhinia reticulata, Terminalia schweinfurthi*** (Bouvier, 1936); ***Berlinia, Combretum*** (Crotch, 1956); ***Brachystegia, Acacia mearnsii, Holarrhena floribunda, Salix*** (Gardiner, 1982).

obscura (Butler) - EUPHORBIACEAE, RHAMNACEAE (Gardiner, 1982).

oyemensis (Rougeot) - ***Robinia pseudoacacia, Laburnum, Fagus*** (Gardiner, 1982).

rectilineata (Sonthonnax) [=***richelmanni*** (Weymer)] (Lemaire, C., pers. comm.) - "Fruit trees", ***Salix*** (Crotch, 1956); "a yellow creeper", ***Diplorhynchus mossambicensis, Uapaca kirkiana, Syzygium owariense*** (Pinhey, 1972).

rubra (Bouvier) - ***Uapaca kirkiana, U. nitida*** (Crotch, 1956); "fruit trees", ***Corylus*** (Gardiner, 1982).

wahlbergi (Boisduval) - ***Psidium, Mangifera indica, Ricinus*** *communis*, *Trema bracteolata* (Platt, 1921); ***Schinus, Quercus*** (Crotch, 1956); ***Maerua, Mangifera,*** *Psidium guajava*, *Prunus persica* (Pinhey, 1972).

ITHOMISA Oberthür

Ithomisa sp. - *Paepalanthus polyanthus* (Crotch, 1956).

LEMAIREIA Nässig & Holloway

luteopeplus aureopeplus Nässig & Holloway - ***Rhus typhina, Liquidambar*** *styraciflua* (Lampe & Nässig, 1989).

LEUCANELLA Lemaire

acutissima (Walker) - ***Rhus*** *laurina* (Wolfe, K.L., pers. comm.).

aspera (C. & R. Felder) ssp. - ***Prosopis algarobilla, Celtis*** *tala*, ***Robinia*** *pseudoacacia*, ***Laburnum, Fagus*** (Gardiner, 1982); ***Ligustrum*** (Naumann, S., pers. comm.).

contempta contempta Lemaire - ***Prunus capuli*** (Gara & Onore, 1989).

leucane (Geyer) - ***Ligustrum*** (Packard, 1914); ***Betula, Pyrus*** *malus*, ***Robinia*** *pseudoacacia* (Villiard, 1969); ***Prunus, Syringa vulgaris*** (Gardiner, 1982).

memusae (Walker) ssp. - ***Ligustrum, Platanus, Acer, Robinia*** *pseudoacacia* (Crotch, 1956); ***Solanum*** *variabile*, ***S. tuberosum*** (Gardiner, 1982).

memusae gardineri Lemaire - ***Ligustrum*** *ovalifolium*, ***Laburnum*** *anagyroides* (Gardiner, 1976).

saturata (Walker) - ***Platanus*** *lindeniana* (Beutelspacher, 1978).

stuarti (Rothschild & Jordan) ssp. - ***Adesmia, Berberis flexuosa,*** *Cassia hookeriana* (Gardiner, 1982).

viridescens (Walker) ssp. - ***Erythrina*** *crista-galli*, ***Platanus, Acer, Ligustrum*** (Gardiner, 1982).

viridescens viridior Lemaire - ***Ligustrum japonicum*** (Probst, H., pers. comm.).

LOBOBUNAEA Packard

acetes (Westwood) - ***Mangifera, Crataegus*** (Gardiner, 1982).

angasana (Westwood) - ***Eugenia cordata, Rhus longifolia,*** *Uapaca kirkiana*, *Apodytes dimidiata* (Platt, 1921); ***Acocanthera, Anisophyllea laurina,*** *Eucalyptus*, *Uapaca nitida* (Pinhey, 1972).

natalensis (Aurivillius) - ***Brachystegia randii*** (Platt, 1921); "fruit trees" (Crotch, 1956).

phaedusa (Drury) - EUPHORBIACEAE, "deciduous trees" (Gardiner, 1982).

saturnus (Fabricius) - ***Anisophyllea laurina,*** *Uapaca nitida*, EUPHORBIACEAE, RHIZOPHORACEAE (Crotch, 1956); ***Acocanthera, Eugenia*** (Gardiner, 1982).

LOEPA Moore

katinka (Westwood) ssp. - ***Cissus, Leea*** (Bouvier, 1936); "vines", *Vitis inconstans*,

Parthenocissus tricuspidata (Crotch, 1956); *Dillenia pentagyna* (Arora & Gupta, 1979); *Crataegus* (Gardiner, 1982).

katinka formosensis Mell - *Tetrastigma formosanum, Ampelopsis brevipedunculata* (Wang, 1988).

oberthuri Leech - *Citrus reticulata* (Zhu & Wang, 1983).

sikkima Moore - *Saurauia* (Holloway, 1987); *Cissus, Leea* (Packard, 1914).

LONOMIA Walker

achelous (Cramer) - *Rhus laurina* (Wolfe, K.L., pers. comm.).

cynira Cramer - *Carpinus betulus, Robinia pseudoacacia, Fagus, Quercus* (Gardiner, 1982).

LUDIA Wallengren

delegorguei (Boisduval) - *Microglossa mesoiloides* (=*M. mespilifolia*), *Senecio deltoideus* (Platt, 1921); *Syringa vulgaris, Tarchonanthus minor, Vernonia natalensis* (Munro, 1946); *Robinia pseudoacacia, Acacia* (Crotch, 1956); *Salvia runcinata, Uapaca kirkiana* (Pinhey, 1972); *Brachylaena discolor, B. rotundata, Tarchonanthus camphoratus, Vernonia oligocephala* (Oberprieler, 1987).

hansali Felder - *Tarchonanthus camphoratus* (Gardiner, 1982).

obscura Aurivillius - ZINGIBERACEAE (Gardiner, 1982).

orinoptena Karsch - *Mikania, Ocimum, Coleus barbatus, C. blumei, Lactuca, Vernonia oligocephala* (Pinhey, 1972); LABIATAE (=LAMIACEAE) (Gardiner, 1982).

MELANOCERA Sonthonnax

menippe (Westwood) - *Ficus cordata, Ochna atropurpurea* (Platt, 1921); *Quercus* (Crotch, 1956); *Bauhinia* (Naumann, S., pers. comm.).

parva Rothschild - *Ochna schweinfurtiana* (Pinhey, 1972).

sufferti (Weymer) - OCHACEAE (Crotch, 1956).

MICRAGONE Walker

ansorgei (Rothschild) - *Berlinia paniculata*, LEGUMINOSAE (=FABACEAE) (Crotch, 1956); *Brachystegia spiciformis, Julbernardia paniculata* (Pinhey, 1972).

cana (Aurivillius) - *Syzygium owariense* (Crotch, 1956); *Pappea* (Pinhey, 1972).

ducorpsi deFleury - *Annona senegalensis* (Crotch, 1956).

nenia Westwood - *Annona senegalensis, Xylopia aethiopica* (Gardiner, 1982).

nubifera Bouvier - *Xylopia aethiopica* (Pinhey, 1972).

MOLIPPA Walker

nibasa Maassen - *Robinia pseudoacacia* (Wolfe, K.L., pers. comm.).

rosea (Druce) - *Platanus* (Simon, L.D., pers. comm.); *Robinia pseudoacacia* (Wolfe, K.L., pers. comm.).

NEORIS Moore

Neoris sp. - *Spiraea, Pyrus* (Crotch, 1956); *Prunus amygdalus, Salix, Ulmus* (Gardiner, 1982).

haraldi Schawerda - *Pyrus malus, P. communis, Prunus persica, Pistacia chinensis* (Zhu & Wang, 1983).

NUDAURELIA Rothschild

alopia (Westwood) - *Trema guineensis* (Gardiner, 1982).

anthina (Karsch) - "False camphor-tree", *Polyscias fulva, Mangifera*, EUPHORBIACEAE (Gardiner, 1982).

bamendana (Schultze) - *Maesa lanceolata* (Gardiner, 1982).

belina (Westwood) - ANACARDIACEAE, *Rhus longifolia* (Crotch, 1956); *Terminalia, Copaifera, Spirocarpus* (=*Medicago*), *Salix* (Gardiner, 1982).

OITICELLA Travassos & Noronha

brevis (Walker) - "A spiny *Mimosa*" (Packard, 1905).

OPODIPHTHERA Wallengren

Opodiphthera sp. - *Elaeocarpus obovatus, Podocarpus elata* (Crotch, 1956).

eucalypti (Scott) - *Schinus molle, Podocarpus spinulosus, Olea paniculata, Geijera salicifolia* (Crotch, 1956); *Eucalyptus blakelyi, E. botryoides, E. bridgesiana, E. cinerea, E. eugenioides, E. fastigiata, E. globulus, E. leucoxylon, E. leucoxylon* var. *macrocarpa, E. macarthuri, E. maculosa, E. macrorhyncha, E. maideni, E. melliodora, E. nitens, E. obliqua, E. paniculata, E. pauciflora, E. regnans, E. saligna, E. viminalis* (White, 1972); *Liquidambar* (Gardiner, 1982); *Eucalyptus gunnii* (Naumann, S., pers. comm.).

helena (White) - *Eucalyptus odorata, E. baxteri, Betula* (Gardiner, 1982).

loranthi (Lucas) - *Viscum, Eucalyptus* (Albaugh, 1986).

ORMISCODES Blanchard

cinnamomea (Feisthamel) - ROSACEAE, *Muehlenbeckia sagittaefolia, Populus, Piper* (Crotch, 1956).

ORTHOGONIOPTILUM Karsch

incana Sonthonnax - *Brachystegia spiciformis* (Pinhey, 1972).

OTHORENE Boisduval

hodeva (Druce) - *Gleditsia triacanthos* (Stone, S.E., pers. obs.).

purpurascens (Schaus) - *Manilkara chicle* (Janzen, 1984a); *Psidium, Terminalia catappa* (Lemaire, 1988).

verana Schaus - *Quercus oleoides* (Janzen, 1984a); *Quercus gambelii* (Stone, S.E., pers. obs.).

PARADIRPHIA Michener

boudinoti (Lemaire & Wolfe) - ***Malus, Prunus, Robinia pseudoacacia*** (Lemaire & Wolfe, 1988).

hoegei (Druce) - ***Platanus lindeniana, Salix chilensis*** (Beutelspacher, 1978); ***Prunus, Robinia pseudoacacia, Pyrus malus*** (Wolfe, K.L., pers. comm.).

lasiocampina C. & R. Felder - ***Quercus*** (Smith, M.J., pers. comm.); ***Pyrus malus, Robinia pseudoacacia*** (Wolfe, K.L., pers. comm.).

semirosea (Walker) - ***Robinia pseudoacacia*** (Lemaire & Wolfe, 1988).

valverdei Lemaire & Wolfe - ***Prunus*** (Lemaire & Wolfe, 1988).

PERISOMENA Walker

caecigena (Kupido) - ***Quercus*** (Packard, 1914); ***Quercus petraea, Q. cerris, Q. robur, Q. pubescens, Q. ilex*** (Rougeot, 1971); ***Populus nigra, P. alba, Pyrus communis*** (Gardiner, 1982); ***Quercus nigra*** (Peigler, R.S., pers. comm.).

PERIPHOBA Hübner

arcaei (Druce) - ***Eugenia salamensis, Cassia biflora, Guazuma ulmifolia, Lysiloma auritum, Spondias mombin, Rourea glabra, Annona purpurea, Calycophyllum candidissimum, Bombacopsis quinatum, Cassia alata, Inga vera, Ardisia revoluta, Astronium graveolens, Hymenaea courbaril, Quercus oleoides, Miconia argentea*** (Janzen, 1984a).

hircia (Cramer) - ***Quercus ilex, Fagus, Crataegus, Prunus, Robinia pseudoacacia, Pyrus malus, Carpinus, Populus nigra*** (Gardiner, 1982); ***Salix babylonica*** (Herbin, D., pers. comm.).

POLYTHYSANA Walker

apollina Felder - ***Persea americana, Liquidambar, Maytenus boaria, Cryptocarya rubra*** (Wolfe, K.L., pers. comm.).

cinerascens Philippi - ***Cryptocarya peumus*** (Crotch, 1956); ***Maytenus boaria, Cryptocarya rubra*** (Wolfe, K.L., pers. comm.).

rubrescens Bloch - ***Maytenus boaria, Cryptocarya rubra*** (Wolfe, K.L., pers. comm.).

PROCITHERONIA Michener

principalis (Walker) - ***Croton floribundus*** (Lemaire, 1988).

PSELAPHELIA Aurivillius

flavivitta (Walker) - ***Aframomum, Trichilia emetica, T. lineata*** (Pinhey, 1972).

gemmifera (Butler) - ***Aframomum*** (Gardiner, 1982).

PSEUDAPHELIA Kirby

apollinaris (Boisduval) - ***Turraea floribunda*** (Platt, 1921); ***Turraea heterophylla*** (Crotch, 1956); ***Combretum, Turraea nilotica*** (Pinhey, 1972).

PSEUDANTHERAEA Weymer

arnobia (Westwood) - *Poga, Terminalia, Uapaca* (Pinhey, 1972).

discrepans (Butler) - *Terminalia, Poga oleosa, Uapaca guineensis, Mangifera, Fagus* (Gardiner, 1982).

PSEUDAUTOMERIS Lemaire

grammivora (Jones) - *Rottboellia compressa* var. *fasciculata* (Gardiner, 1982).

irene (Cramer) ssp. - *Passiflora quadrangulare* (Gardiner, 1982).

ophthalmica (Moore) - *Iris* (Gardiner, 1982).

salmonea (Cramer) - *Citrus, Passiflora quadrangulare* (Gardiner, 1982).

PSEUDIMBRASIA Rougeot

deyrollei (Thomson) - *Erythrophleum guineense*, EUPHORBIACEAE, MORACEAE, RUBIACEAE (Pinhey, 1972); *Quercus* (Gardiner, 1982).

PSEUDOBUNAEA Bouvier

Pseudobunaea sp. - *Annona senegalensis, Bauhinia reticulata* (Bouvier, 1936).

alinda (Drury) - CAESALPINIACEAE (=FABACEAE: CAESALPINIOIDEAE), *Fagus* (Gardiner, 1982).

callista (Jordan) - *Brachystegia, Julbernardia* (Pinhey, 1972).

cleopatra (Aurivillius) - *Alchornea cordifolia, Bridelia grandis, Croton oligandrum*, EUPHORBIACEAE (Gardiner, 1982).

epithyrena (Maassen & Weymer) - *Brachystegia randi* (Platt, 1921); *Pterocarpus, Uapaca, Berlinia* (Crotch, 1956).

illustris (Weymer) - *Rosa* (Gardiner, 1982).

irius (Fabricius) - *Populus deltoides* (Hepburn et al., 1966); *Bauhinia, Brachystegia spiciformis, ?Julbernardia paniculata, Pterocarpus delevoyi, Uapaca nitida* (Pinhey, 1972); *Pinus patula* (Van den Burg & Van den Burg, 1973).

pallens (Sonthonnax) - *Berlinia paniculata, Bauhinia reticulata, Uapaca nitida, Fagus* (Gardiner, 1982).

tyrrhena (Westwood) - *Celtis kraussiana, Albizia fastigiata* (Packard, 1914); *Ekebergia meyeri, Trema bracteolata* (Platt, 1921); *Grumilea capensis* (Bouvier, 1936); *Ulmus, Fagus, Celtis pallida*, RUBIACEAE, BURSERACEAE, LECYTHIDACEAE (Gardiner, 1982).

PSEUDODIRPHIA Bouvier

eumedide Stoll - *Fagus sylvatica, Ulmus campestris, Laburnum anagyroides, Robinia pseudoacacia, ?Quercus ilex, Crataegus oxyacantha* (Gardiner, 1965).

eumedidoides (Vuillot) - *Salix babylonica* (Herbin, D., pers. comm.); *Robinia pseudoacacia* (Wolfe, K.L., pers. comm.).

PSILOPYGIDA Michener

crispula (Dognin) - *Acacia farnesiana, Gleditsia triacanthos* (Lemaire, 1988).

PTILOSCOLA Michener

dargei (Lemaire) - ***Acacia tenuifolia*** (Janzen, 1984a).

RHESCYNTIS Hübner

hippodamia (Cramer) ssp. - ***Virola bicuhyba, V. ucuuba*** (Gardiner, 1982); ***Ligustrum*** (Simon, L.D., pers. comm.).

RHODINIA Staudinger

fugax Butler - ***Phellodendron amurense, Fagus*** (Bouvier, 1936); ***Quercus, Salix*** (Crotch, 1956); ***Acer, Quercus acutissima, Q. dentata, Q. serrata, Prunus, Zelkova serrata*** (Inoue *et al.*, 1982); ***Hevea brasiliensis, Cyclobalanopsis glauca*** (Zhu & Wang, 1983); ***Castanea, Ilex verticillata, Celtis pallida, Platanus*** (Albaugh, 1986).

jankowskii Oberthür - ***Phellodendron amurense, Fagus*** (Bouvier, 1936).

newara (Moore) - ***Salix, Juglans regia, Acer, Crataegus*** (Crotch, 1956); ***Acer saccharum, Salix babylonica, S. elegans*** (Arora & Gupta, 1979).

verecunda Inoue - ***Quercus acutissima, Q. aliena, Celtis sinensis, Cyclobalanopsis glauca*** (Wang, 1988).

ROHANIELLA Bouvier

pygmaea (Maassen & Weymer) - ***Burkea africana*** (Pinhey, 1972).

ROTHSCHILDIA Grote

arethusa (Walker) ssp. - ***Ailanthus altissima, Ligustrum, Quercus, Syringa vulgaris*** (Crotch, 1956).

aurota (Cramer) ssp. - ***Ricinus communis*** (Bouvier, 1936); ***Ailanthus altissima, Ligustrum*** (Crotch, 1956).

aurota speculifera (Walker) - ***Prunus, Syringa vulgaris, Cephalanthus, Fraxinus, Ligustrum*** (Albaugh, 1986).

condor (Staudinger) - ***Baccharis sculpta*** (Bouvier, 1936).

cincta (Tepper) ssp. - ***Quercus, Prunus serotina*** (Sblendorio, D.C., pers. comm.); ***Jatropha, Schinus*** (Smith, M.J., pers. comm.); ***Ligustrum*** (Tuttle, J.P., pers. comm.).

erycina (Shaw) ssp. - AURANTEACEAE (=RUTACEAE) (Bouvier, 1936); ***Exostema mexicanum, Coutarea hexandra, Ligustrum*** (Janzen, 1984a); ***Ailanthus altissima*** (Wolfe, K.L., pers. comm.).

hesperus (Linnaeus) ssp. - ***Prunus armeniaca, Ligustrum, Ailanthus altissima*** (Wolfe, K.L., pers. comm.).

jacobaeae (Walker) - ***Ligustrum*** (Crotch, 1956); ***Syringa vulgaris*** (Gardiner, 1982); ***Prunus serotina*** (Purdum, D.A., pers. comm.); ***Prunus virginiana*** (Stone, S.E., pers. obs.).

jorulla (Westwood) ssp. - ***Fraxinus, Prunus serotina*** (Packard, 1914); ***Syringa vulgaris, Ailanthus altissima, Quercus*** (Crotch, 1956); ***Mimosa*** (Collins & Weast, 1961); ***Jatropha eurcas*** (Comstock & Garcia, 1964); ***Ligustrum, Salix, Sapium***

biloculare (Ferguson, 1971-72); *Salix discolor* (Williams, B.D., pers. comm.).

lebeau (Guérin-Méneville) ssp. - *Ligustrum, Salix, Fraxinus, Prunus, Citrus, Zanthoxylum fagara* (Gardiner, 1982); *Zuelania guidonia* (Janzen, 1982); *Exostema mexicanum, Spondias mombin, S. purpurea, Casearia corymbosa, Zuelania guidonia, Xanthoxylum* (=*Zanthoxylum*) *setulosum* (Janzen, 1984a); *Rhus laurina, Schinus terebinthifolius, Ailanthus altissima* (Wolfe, K.L., pers. comm.).

lebeau aroma Schaus - *Prunus serotina* (Purdum, D.A., pers. comm.).

lebeau forbesi Benjamin - *Ligustrum, Salix* (Benjamin, 1934); *Fraxinus* (Collins & Weast, 1961); *Citrus, Zanthoxylum fagara* (Ferguson, 1971-72); *Salix interior, Prunus persica, P. serotina, Syringa vulgaris* (Tuttle, J.P., pers. comm.); *Salix discolor* (Williams, B.D., pers. comm.).

maurus (Burmeister) - ANACARDIACEAE, BIGNONIACEAE, EUPHORBIACEAE, *Croton* (Bouvier, 1936).

maurusius Schreiter - See ***Rothschildia maurus*** (Burmeister) (Lemaire, 1978: 77).

orizaba (Westwood) ssp. - *Fraxinus americana* (Packard, 1914); *Ligustrum, Prunus* (Crotch, 1956); *Prunus armeniaca, Salix, Schinus molle, Syringa vulgaris, Cephalanthus, Sassafras albidum* (Villiard, 1969); *Ligustrum japonicum* (Ferguson, 1971-72); *Prunus virginiana, P. serotina* (Sblendorio, D.C., pers. comm.); *Rhus laurina* (Wolfe, K.L., pers. comm.).

roxana Schaus - *Prunus serotina* (Purdum, D.A., pers. comm.); *Ligustrum* (Stone, S.E., pers. obs.).

splendidus (Kirby) - See ***Rothschildia erycina*** (Shaw) ssp. (Lemaire, 1978: 42).

tucumana (Dognin) - *Eupatorium buniifolium* (Bouvier, 1936).

SALASSA Moore

iris Jordan - *Prunus* (Allen, M.G., pers. comm.).

SAMIA Hübner

cynthia (Drury) ssp. - *Ailanthus altissima* (Holland, 1903); *Cornus, Ricinus communis, Prunus* (Packard, 1914); *Ligustrum, Syringa vulgaris* (Crotch, 1956); *Forsythia* (Rougeot, 1971); *Fraxinus, Sassafras albidum, Liriodendron tulipifera, Lindera benzoin, Liquidambar styraciflua, Platanus, Tilia, Juglans cinerea, Ilex* (Ferguson, 1971-72); *Althaea rosa, Apium graveolens, Azadirachta indica, Berberis, Carica papaya, Carpinus betulus, Cassia fistula, Coriaria nepalensis, Euodia fraxinifolia, Gmelina arborea, Heteropanax fragrans, Jatropha curcas, Juglans regia, Lawsonia alba, Manihot utilissima, Plumeria acuminata, Pyrus malus, Zanthoxylum acanthopodium, Z. hostile, Z. alatum, Ziziphus jujuba* (Arora & Gupta, 1979); *Lagerstroemia indica* (Barlow, 1982); *Ilex chinensis, Michelia figo, Firmiana simplex* (Zhu & Wang, 1983); *Salix* (Albaugh, 1986); *Rhus typhina* (Sblendorio, D.C., pers. comm.); *Prunus serotina* (Williams, B.D., pers. comm.).

cynthia ricini Boisduval - *Manihot esculenta, Ricinus communis* (Zhu & Wang, 1983).

cynthia walkeri C. & R. Felder - ***Ricinus communis, Lindera megaphylla, Sapium discolor, S. sebiferum, Cinnamomum camphora, Euscaphis japonica, Glochidion hongkongense*** (Wang, 1988).

SATURNIA Schrank

albofasciata (J. W. Johnson) - ***Cercocarpus betuloides, Ceanothus cuneatus, C. cordulatus, C. leucodermis, C. spinosa, Fremontia californica*** (Hogue *et al.*, 1965); ***Cercocarpus montanus*** (Stone, S.E., pers. obs.).

atlantica H. Lucas - ***Fraxinus berlandierana*** (Rougeot, 1971); ***Olea europaea, Prunus armeniaca*** (Gardiner, 1982); ***Alnus, Rhamnus, Rosa, Syringa vulgaris*** (Rougeot & Viette, 1983); ***Pyrus malus*** (Stone, S.E., pers. obs.).

cephalariae Christoph - ***Cephalaria procera*** (Rougeot, 1971).

lindia Moore - ***Corylus avellana, Betula pendula, Pyrus calleryana, Malus, Salix, Populus deltoides, Acer, Quercus robur, Juglans regia, Prunus avium, P. padus*** (Nässig, 1983); ***Acer saccharinum*** (Peigler, R.S., pers. comm.).

mendocino Behrens - ***Arctostaphylos tomentosa*** (Edwards, 1880)[5]; ***Arbutus menziesii*** (Tilden, 1945); ?***Rhus trilobata***, ?***R. integrifolia, Ceanothus cuneatus*** (Ferguson, 1971-72); ***Arctostaphylos glauca*** (Tietz, 1972); ***Rhus typhina, Arctostaphylos uva-ursi, A. alpinus*** (Naumann, S., pers. comm.).

pavonia (Linnaeus) ssp. - ***Spiraea alba*** (Villiard, 1969); ***Potentilla, Hippophae rhamnoides, Fagus, Fragaria, Myrica gale, Quercus, Rumex, Sambucus, Sorbus, Ulmus, Vaccinium myrtillus, Juglans*** (Rougeot, 1971); ***Rosa, Rubus, Crataegus, Pyrus malus, Crataegus, Prunus spinosa, Calluna vulgaris, Salix, Betula, Lythrum salicaria, Alnus, Fraxinus, Humulus*** (Albaugh, 1986); ***Arbutus unedo, Pistacia terebinthus, Cornus*** (Lemaire, C., pers. comm.); ***Prunus domestica, Salix babylonica, Rubus idaeus, Pyrus communis*** (Naumann, S., pers. comm.); ***Erica, Prunus serotina*** (Peigler, R.S., pers. comm.); ***Schinus terebinthifolius*** (Smith, M.J., pers. comm.); ***Prunus virginiana*** (Stone, S.E., pers. obs.).

pyri (Denis & Schiffermüller) - "Fruit trees", ***Pyrus malus, Prunus*** (Crotch, 1956); ***Aesculus hippocastanum*** (Villiard, 1969); ***Olea europaea, Fraxinus excelsior, Acer pseudoplatanus, Alnus glutinosa, Amygdalus communis, Cydonia, Juglans regia, Prunus spinosa, Ulmus campestris, Salix, Syringa, Platanus*** (Rougeot, 1971); ***Castanea, Populus*** (Gardiner, 1982); ***Fraxinus, Ligustrum, Betula, Pyrus communis, Humulus, Tilia, Prunus amygdalus*** (Albaugh, 1986); ***Prunus armeniaca*** (Stone, S.E., pers. obs.).

---

[5] Packard (1914: 151, footnote) and Ferguson (1971-72: 181) incorrectly cite Edwards (Proc. Calif. Acad. Sci. 1878, January) as the reference for the larval description and foodplant record for *S. mendocino*. According to Mr. Greg Lym, Assistant Librarian, User Services, California Academy of Sciences, there were no proceedings published for the Academy in 1878. The correct citation is provided in Literature Consulted here.

spini (Denis & Schiffermüller) - *Cephalaria procera* (Rougeot, 1971); *Prunus spinosa, Pyrus malus, Ulmus campestris, Crataegus oxyacantha, Alnus glutinosa, Rosa, Rubus, Salix, Populus* (Gardiner, 1982); *Salix babylonica* (Naumann, S., pers. comm.); *Prunus* (Wolfe, K.L., pers. comm.).

walterorum Hogue & J. W. Johnson - *Arctostaphylos, Rhus laurina* (Sala & Hogue, 1958); *Rhus integrifolia* (Tuskes, 1974); *Arbutus unedo* (Lemaire, C., pers. comm.); *Rhus trilobata* (Stone, S.E., pers. obs); *Schinus terebinthifolius* (Wolfe, K.L., pers. comm.).

SCHAUSIELLA Bouvier

arpi (Schaus) - *Hymenaea altissima* (Lemaire, 1988).

santarosensis Lemaire - *Hymenaea courbaril* (Janzen, 1984a).

SYSSPHINX Hübner

albolineata (Grote & Robinson) - *Acacia angustissima* (Tuskes, 1985a); *Acacia baileyana, Robinia pseudoacacia* (Wolfe, K.L., pers. comm.).

bicolor (Harris) - *Gymnocladus canadensis, Gleditsia triacanthos* (Packard, 1893); *Gymnocladus dioica* (Holland, 1903); *Hymenaea altissima* (Forbes, 1923); *Quercus* (Tietz, 1972); *Robinia pseudoacacia, ?Ligustrum* (Gardiner, 1982).

bisecta (Lintner) - *Gleditsia triacanthos, Gymnocladus canadensis* (Eyer & Menke, 1914); *Robinia pseudoacacia* (Lemaire, 1988).

blanchardi (Ferguson) - *Acacia flexicaulis* (Ferguson, 1971-72); *Gleditsia triacanthos* (Tuttle, J.P., pers. comm.).

colla Dyar ssp. - *Pithecellobium saman* (Janzen, 1984a).

gomezi Lemaire - *Acacia baileyana, Robinia pseudoacacia* (Wolfe, K.L., pers. comm.).

heiligbrodti (Harvey) - *Prosopis glandulosa, Acacia farnesiana, A. rigidula* (Blanchard, 1964); *Cercidium floridum, C. microphyllum* (Tietz, 1972); *Gleditsia triacanthos* (Tuttle, J.P., pers. comm.).

hubbardi Dyar - *Parkinsonia microphyllum, P. torreyana, Prosopis* (Kunzé, 1906); *Gleditsia triacanthos* (Lemaire, 1988); *Acacia* (Smith, M.J., pers. comm.); *Prosopis juliflora* (Tuttle, J.P., pers. comm.); *Robinia pseudoacacia* (Williams, B.D., pers. comm.).

jason (Boisduval) ssp. - See ***Adeloneivaia jason*** (Boisduval) ssp., (Lemaire, 1988: 369).

mexicana (Boisduval) - *Acacia collinsii, A. cornigera* (Janzen, 1984a).

molina (Cramer) - "A high forest *Mimosa*" (Packard, 1905); *Pithecellobium saman, Cassia grandis, Albizia adinocephala* (Janzen, 1984a); *Robinia pseudoacacia, Salix capraea* (Lemaire, 1988); *Gleditsia triacanthos* (Smith, M.J., pers. comm.); *Robinia, Acacia baileyana* (Wolfe, K.L., pers. comm.).

montana (Packard) - *Acacia farnesiana, Gleditsia triacanthos, Cassia emarginata, Haematoxylum brasalita* (Tuskes, 1985a); *Albizia* (Wilson, M.F., pers. comm.); *Robinia pseudoacacia* (Wolfe, K.L., pers. comm.).

petersii Peters - See ***Syssphinx molina*** (Cramer) (Lemaire, 1988: 243).
quadrilineata (Grote & Robinson) ssp. - *Acacia flexicaulis, Parkinsonia microphyllum, Prosopis juliflora* (Packard, 1905); *Gleditsia triacanthos* (Heitzman, 1961); *Acacia collinsii* (Janzen, 1984a); *Pithecellobium saman* (Janzen, 1985); *Acacia dealbata* (Herbin, D., pers. comm.); *Acacia baileyana* (Wolfe, K.L., pers. comm.).
raspa (Boisduval) - *Acacia angustissima, Gleditsia texana* (Tuskes, 1985a).

TAGOROPSIS Felder
flavinata (Walker) - *Schmidelia africana* (=*Allophyllus melanocarpus*) (Platt, 1921); *Allophyllus africanus* (Bouvier, 1936).
hanningtoni Butler - *Allophyllus africanus* "and other SAPINDACEAE" (Crotch, 1956).
songeana Strand - *Schmidelia africana* (=*Allophyllus melanocarpus*) (Pinhey, 1972).
subocellata (Butler) - *Agauria salicifolia*, EUPHORBIACEAE (Griveaud, 1961).

TITAEA Hübner
tamerlan (Maassen) ssp. - *Bombax* (Gardiner, 1982); *Bombacopsis quinatum* (Janzen, 1984a); *Tilia platyphyllos* (Herbin, D., pers. comm.).

UROTA Westwood
sinope Westwood - *Sclerocarya caffra* (Platt, 1921); *Erythrina abyssinica, E. caffra* (Crotch, 1956); *Erythrina tomentosa* (Pinhey, 1972); *Laburnum, Robinia pseudoacacia* (Gardiner, 1982).

USTA Wallengren
terpsichore (Maassen & Weymer) - *Commiphora caryaefolia* (Platt, 1921); BURSERACEAE (Crotch, 1956); *Schinus molle, Melia azedarach, Sclerocarya caffra* (Pinhey, 1972).

VEGETIA Jordan
ducalis Jordan - *Eriocephalus umbellatus, Eriosema* (Pinhey, 1972); *Eriocephalus africanus, E. ericoides* (Oberprieler, 1987); *Artemisia californica* (Wolfe, K.L., pers. comm.).
grimmia (Geyer) - *Eriocephalus racemosus* (Oberprieler, 1987).
dewitzi (Maassen & Weymer) - *Eriocephalus? ericoides* (Oberprieler, 1987).

FOODPLANTS OF SATURNIID HYBRIDS

Attacus atlas ♂ X *A. lorquinii* ♀
Ligustrum, Ailanthus, & Berberis (Peigler, 1989).

Actias luna luna ♂ X *A. truncatipennis* ♀
Juglans (Wolfe, K. L., pers. comm.).

Actias truncatipennis ♂ X *A. luna luna* ♀
Juglans (Wolfe, K. L., pers. comm.).

Anisota virginiensis ♂ X *A. discolor* ♀
Probably *Quercus* (Riotte & Peigler, 1981).

Anisota peigleri ♂ X *A. consularis* ♀
Probably *Quercus* (Riotte & Peigler, 1981).

Anisota senatoria ♂ X *A. oslari* ♀
Quercus coccinea (Peigler & Williams, 1984).

Automeris louisiana ♂ X *A. io io* ♀
Celtis (Tuskes, P. M., pers. comm.).

Callosamia promethea ♂ X *C. securifera* ♀
Liriodendron tulipifera (Peigler, 1977).

Callosamia angulifera ♂ X *C. promethea* ♀
Liriodendron tulipifera (Peigler, 1977).

Callosamia angulifera ♂ X *C. securifera* ♀
Liriodendron tulipifera, Magnolia virginiana (Peigler, 1977).

Callosamia securifera ♂ X *C. promethea* ♀
Liriodendron tulipifera, Magnolia virginiana (Peigler, 1981).

Callosamia [*angulifera* ♂ X (*angulifera* ♂ X *securifera* ♀) ♀] ♂ X *C. securifera* ♀
Liriodendron tulipifera (Peigler, 1981).

Callosamia angulifera ♂ X *C.* (*angulifera* ♂ X *securifera* ♀) ♀
Liriodendron tulipifera (Peigler, pers. comm.).

Callosamia (angulifera ♂ X *securifera* ♀) ♂ X *C. securifera* ♀
Magnolia virginiana until last two instars then transferred to *Liriodendron tulipifera* (Peigler, 1977).

Callosamia (securifera ♂ X *angulifera* ♀) ♂
X *C.* [*angulifera* ♂ X (*angulifera* ♂ X *securifera* ♀) ♀] ♀
Liriodendron tulipifera (Peigler, 1977).

Callosamia (securifera ♂ X *angulifera* ♀) ♂ X (*angulifera* ♂ X *securifera* ♀) ♀
Liriodendron tulipifera (Peigler, 1977).

Callosamia securifera ♂ X C. (*securifera* ♂ X *angulifera* ♀) ♀
Magnolia virginiana (Peigler, 1977).

Callosamia securifera ♂ X *C. angulifera* ♀
Liriodendron tulipifera (Peigler, 1977).

Callosamia angulifera ♂ X *Hyalophora cecropia* ♀
Liriodendron tulipifera (Carr, 1984).

Callosamia (angulifera ♂ X *securifera* ♀) F_2
Liriodendron tulipifera (Peigler, 1977).

Callosamia angulifera ♂ X *Samia cynthia* ♀
Liriodendron tulipifera (Peigler, 1978).

Citheronia regalis ♂ X *C. splendens* ssp. ♀
Liquidambar styraciflua (Simon, L. D., pers. comm.); *Arctostaphylos* (Stone, S. E., pers. obs.).

Graellsia isabellae ♂ X *Actias luna luna* ♀
Juglans (Vuattoux, R., pers. comm.).

Hemileuca artemis ♂ X *H. maia* ♀
Salix (Carr, T. W., pers. comm.).

Hemileuca maia ♂ X *H. artemis* ♀
Salix (Carr, T. W., pers. comm.).

Hemileuca maia ♂ X *H. lucina* ♀ F_1
Salix (Carr, T. W., pers. comm.).

Hemileuca maia ♂ X *H. lucina* ♀ F_2
Salix (Carr, T. W., pers. comm.).

Hemileuca griffini ♂ X *H. chinatiensis* ♀
Amelanchier utahensis (Stone, S. E., pers. obs.).

Hemileuca nuttalli ♂ X *H. eglanterina shastaensis* ♀
Symphoricarpos albus (Stone, S. E., pers. obs.).

Hemileuca lucina ♂ X *Hemileuca nevadensis* ♀
Salix gracilis, S. ?nigra (Peigler & Williams, 1984).

Hyalophora cecropia ♂ X *H. columbia gloveri* ♀
Sambucus, Prunus virginiana (Stone, S. E., pers. obs.).

Hyalophora euryalus euryalus ♂ X *H. cecropia* ♀
Sambucus (Stone, S. E., pers. obs.).

Hyalophora columbia gloveri ♂ X *H. cecropia* ♀
Sambucus, Prunus virginiana (Stone, S. E., pers. obs.).

Hyalophora cecropia ♂ X *H. euryalus euryalus* ♀
Sambucus (Stone S. E., pers. obs.).

Hyalophora cecropia ♂ X *H. columbia columbia* ♀
Prunus serotina (Packard, 1914); *Prunus virginiana* (Stone, S. E., pers. obs.).

Hyalophora columbia columbia ♂ X *H. columbia gloveri* ♀
Prunus virginiana (Stone, S. E., pers. obs.).

Hyalophora columbia columbia ♂ X *H. euryalus euryalus* ♀
Larix laricina, Salix (Köhalmi & Moens, 1975); *Prunus virginiana* (Stone, S. E., pers. obs.).

Hyalophora columbia gloveri ♂ X *H. columbia columbia* ♀
Prunus virginiana (Stone, S. E., pers. obs.).

Hyalophora euryalus euryalus ♂ X *H. columbia gloveri* ♀
Prunus virginiana (Stone, S. E., pers. obs.).

Hyalophora columbia columbia ♂ X *H. cecropia* ♀
Larix laricina, Salix (Köhalmi & Moens, 1975); *Prunus virginiana* (Stone, S. E., pers. obs.).

Hyalophora columbia nokomis ♂ X *H. cecropia* ♀
Prunus serotina or *Sambucus* or *Salix* or *Alnus* (Sweadner, 1937).

Hyalophora columbia gloveri ♂ X *H. euryalus euryalus* ♀
Prunus virginiana (Stone, S. E., pers. obs.).

Hyalophora euryalus euryalus ♂ X *H. columbia columbia* ♀
Prunus virginiana (Stone, S. E., pers. obs.).

Hyalophora (*cecropia* ♂ X *columbia gloveri* ♀) ♂ X *H. euryalus euryalus* ♀
Salix babylonica, Prunus serotina (Wright, 1971).

Hyalophora [(*cecropia* ♂ X *columbia gloveri* ♀) ♂ X *H. euryalus euryalus* ♀] ♂ X *H. columbia columbia* ♀
Salix babylonica, Prunus serotina, Larix americana (Wright, 1971).

Hyalophora (*cecropia* ♂ X *columbia gloveri* ♀) ♂ X *H. cecropia* ♀
Prunus serotina or *Sambucus* or *Salix* or *Alnus* (Sweadner, 1937).

Hyalophora cecropia ♂ X (*H. cecropia* ♂ X *H. columbia gloveri* ♀) ♀
Prunus serotina or *Sambucus* or *Salix* or *Alnus* (Sweadner, 1937).

Hyalophora columbia gloveri ♂ X (*cecropia* ♂ X *H. columbia gloveri* ♀) ♀
Prunus serotina or *Sambucus* or *Salix* or *Alnus* (Sweadner, 1937)

Hyalophora cecropia ♂ X *H. columbia gloveri* ♀ F_2
Prunus serotina or *Sambucus* or *Salix* or *Alnus* (Sweadner, 1937).

Hyalophora (*columbia columbia* ♂ X *cecropia* ♀) ♂ X *H. cecropia* ♀
Prunus serotina or *Sambucus* or *Salix* or *Alnus* (Sweadner, 1937).

Hyalophora (*columbia columbia* ♂ X *cecropia* ♀) ♂ X *H. cecropia* ♀
Prunus serotina or *Sambucus* or *Salix* or *Alnus* (Sweadner, 1937).

Saturnia walterorum ♂ X *S. mendocino* ♀
Arctostaphylos (Tuskes & Collins, 1981).

Saturnia mendocino ♂ X *S. walterorum* ♀
Arctostaphylos (Tuskes & Collins, 1981).

Saturnia (*walterorum* ♂ X *mendocino* ♀) ♂ X *S. mendocino* ♀
Arctostaphylos (Tuskes & Collins, 1981).

Saturnia (*walterorum* ♂ X *mendocino* ♀) ♂ X *S. walterorum* ♀
Arctostaphylos (Tuskes & Collins, 1981).

Saturnia (*mendocino* ♂ X *walterorum* ♀) F_1
Arctostaphylos (Tuskes & Collins, 1981).

Saturnia (*walterorum* ♂ X *mendocino* ♀) F_1
Arctostaphylos (Tuskes & Collins, 1981).

Saturnia ([*walterorum* ♂ X *mendocino* ♀] ♂ X *S. walterorum* ♀) F_2
Arctostaphylos (Tuskes & Collins, 1981).

Syssphinx bicolor ♂ X *S. hubbardi* ♀
Gleditsia triacanthos (Carr, T. W., pers. comm.).

PART 2
SATURNIID FOODPLANT GENERA AND SPECIES

ABIES (PINACEAE)

balsamea - *Automeris io* ssp., *Hyalophora columbia* ssp.
grandis - *Eacles imperialis* ssp.

ACACIA (FABACEAE: MIMOSOIDEAE)

sp. unspecified - *Caligula boisduvali* ssp., *Gynanisa ata, G. nigra, Hemileuca chinatiensis, H. rubridorsa, Imbrasia ertli, Ludia delegorguei, Syssphinx albolineata, S. hubbardi.*
angustissima - *Syssphinx albolineata, S. raspa.*
baileyana - *Hylesia lineata, Syssphinx albolineata, S. gomezi, S. molina, S. quadrilineata* ssp.
collinsii - *Syssphinx mexicana, S. quadrilineata* ssp.
cornigera - *Syssphinx mexicana.*
cyanophylla - *Imbrasia cytherea.*
cyclopis - *Imbrasia cytherea.*
dealbata - *Adeloneivaia boisduvali, Syssphinx quadrilineata.*
farnesiana - *Psilopygida crispula, Syssphinx heiligbrodti, S. montana.*
flexicaulis - *Syssphinx blanchardi, S. quadrilineata* ssp.
greggi - *Hemileuca burnsi, H. tricolor.*
karroo - *Gonimbrasia tyrrhea* ssp., *Gynanisa maja* ssp.
mearnsii - *Imbrasia nictitans.*
mollissima - *Gonimbrasia tyrrhea* ssp., *Gynanisa maja* ssp., *Heniocha apollonia, H. dyops, H. marnois, Imbrasia cytherea.*
nilotica - *Cirina forda.*
rigidula - *Syssphinx heiligbrodti.*
saligna - *Imbrasia cytherea.*
sieberiana - *Goodia kuntzei.*
tenuifolia - *Hylesia lineata, Ptiloscola dargei.*

ACALYPHA (EUPHORBIACEAE)

sp. unspecified - *Citheronia laocoon.*

ACER (ACERACEAE)

sp. unspecified - *Actias artemis, A. luna* ssp., *Aglia tau* ssp., *Antheraea polyphemus* ssp., *Automeris coresus, A. io* ssp., *Caligula boisduvali* ssp., *C. japonica* ssp., *Callosamia promethea, Eacles imperialis* ssp., *Hemileuca eglanterina* ssp., *?Hyalophora columbia columbia, H. columbia gloveri, Leucanella memusae* ssp., *L. viridescens* ssp., *Rhodinia fugax, R. newara, Saturnia lindia.*
caudatum - *Caligula zuleika.*

circinatum - *Hemileuca eglanterina* ssp.
dasycarpum - *Hyalophora cecropia.*
monspessulanum - *Dryocampa rubicunda rubicunda.*
negundo - *Anisota oslari, Antheraea polyphemus* ssp., *Automeris io io, Dryocampa rubicunda* ssp., *Eacles imperialis* ssp., *Hyalophora cecropia, H. columbia gloveri.*
pensylvanicum - *Antheraea polyphemus* ssp.
plantanoides - *Hyalophora cecropia.*
pseudoplatanus - *Aglia tau* ssp., *Eacles imperialis* ssp., *Saturnia pyri.*
rubrum - *Actias luna* ssp., *Anisota oslari, Antheraea polyphemus* ssp., *Automeris io* ssp., *Dryocampa rubicunda* ssp., *D. rubicunda rubicunda, Eacles imperialis* ssp., *Hyalophora cecropia.*
saccharinum - *Anisota oslari, Antheraea polyphemus* ssp., *Eacles imperialis* ssp., *E. imperialis imperialis, Dryocampa rubicunda* ssp., *D. rubicunda alba, D. rubicunda rubicunda, Hyalophora cecropia, Saturnia lindia.*
saccharum - *Actias luna* ssp., *Dryocampa rubicunda rubicunda, Eacles imperialis* ssp., *Rhodinia newara.*
spicatum - *Dryocampa rubicunda rubicunda, Hyalophora cecropia.*

ACOCANTHERA (APOCYNACEAE)
sp. unspecified - *Imbrasia krucki, Lobobunaea angasana, L. saturnus.*

ACTINODAPHNE (LAURACEAE)
sikkimensis - *Caligula zuleika.*

ADENOSTOMA (ROSACEAE)
fasciculatum - *Hemileuca eglanterina* ssp.

ADESMIA (FABACEAE: FABOIDEAE)
sp. unspecified - *Leucanella stuarti* ssp.

ADINANDRA (THEACEAE)
dumosa - *Actias maenas.*

AESCULUS (HIPPOCASTANACEAE)
hippocastanum - *Caligula japonica* ssp., *Eacles imperialis* ssp., *Saturnia pyri.*

AFRAMOMUM (ZINGIBERACEAE)
sp. unspecified - *Goodia lunata, G. sentosa, Pselaphelia gemmifera, P. flavivitta.*

AGAURIA (ERICACEAE)
salicifolia - *Tagoropsis subocellata.*

AGROPYRON (POACEAE)
smithii - *Hemileuca oliviae.*

AILANTHUS (SIMAROUBACEAE)
altissima - *Archaeoattacus edwardsii, Attacus atlas, Rothschildia arethusa* ssp., *R. aurota* ssp., *R. erycina* ssp., *R. hesperus* ssp., *R. jorulla* ssp., *R. lebeau* ssp., *Samia cynthia* ssp.

ALBIZIA (FABACEAE: MIMOSOIDEAE)
sp. unspecified - *Aurivillius triramis, Dysdaemonia fosteri, Eustera argiphontes, Gonimbrasia tyrrhea* ssp., *Syssphinx montana.*
adinocephala - *Syssphinx molina.*
fastigiata - *Aurivillius arata, Pseudobunaea tyrrhena.*
gummifera - *Aurivillius arata.*
versicolor - *Cirina forda.*

ALCHORNEA (EUPHORBIACEAE)
cordifolia - *Pseudobunaea cleopatra.*

ALEURITES (EUPHORBIACEAE)
fordii - *Citheronia laocoon, Imbrasia dione.*
montana - *Attacus atlas, Imbrasia dione.*

ALLOPHYLUS (SAPINDACEAE)
africanus - *Tagoropsis flavinata, T. hanningtoni.*
melanocarpus - *Tagoropsis flavinata, T. songeana.*
occidentalis - *Hylesia lineata.*

ALNUS (BETULACEAE)
sp. unspecified - *Actias luna* ssp., *Antheraea polyphemus* ssp., *Eacles imperialis* ssp., *E. imperialis imperialis, Hyalophora cecropia, H. columbia gloveri, Saturnia atlantica, S. pavonia ssp.*
cremastogyne - *Actias selene ningpoana, Aglia tau* ssp.
firma - *Actias artemis.*
formosana - *Actias selene ningpoana.*
glutinosa - *Actias selene* ssp., *Hyalophora cecropia, Saturnia pyri, S. spini.*
incana - *Antheraea polyphemus* ssp., *Hyalophora cecropia.*
japonica - *Actias selene gnoma, Aglia tau* ssp.
rubra - *Hyalophora euryalus euryalus.*
serrulata - *Antheraea polyphemus* ssp., *Eacles imperialis* ssp., *Hyalophora cecropia.*

ALOE (LILIACEAE)
sp. unspecified - *Cirina forda.*

ALSTONIA (APOCYNACEAE)
scholaris - *Attacus atlas*.

ALTHAEA (MALVACEAE)
rosa - *Samia cynthia* ssp.

AMELANCHIER (ROSACEAE)
sp. unspecified - *Hemileuca griffini, H. eglanterina* ssp.
alnifolia - *Hyalophora euryalus euryalus*.
canadensis - *Antheraea polyphemus* ssp., *Hyalophora cecropia*.
florida - *Hyalophora euryalus euryalus*.

AMOMUM (ZINGIBERACEAE)
sp. unspecified - *Goodia lunata, G. sentosa*.

AMORPHA (FABACEAE: FABOIDEAE)
sp. unspecified - *Automeris io io*.
fruticosa - *Automeris io* ssp., *A. io lilith*.

AMPELOPSIS (VITACEAE)
brevipedunculata - *Loepa katinka formosensis*.

AMYGDALUS (ROSACEAE)
communis - *Citheronia brissotii* ssp.

ANACARDIUM (ANACARDIACEAE)
sp. unspecified - *Hyperchiria nausica*.
occidentale - *Citheronia laocoon, C. armata armata, Cricula trifenestrata, Imbrasia dione*.

ANDROMEDA (ERICACEAE)
glaucophylla - *Actias selene* ssp.
ovalifolia - *Actias selene* ssp., *Caligula thibeta* ssp.

ANDROPOGON (POACEAE)
hallii - *Hemileuca oliviae*.

ANISOPHYLLEA (RHIZOPHORACEAE)
laurina - *Lobobunaea angasana, L. saturnus*.

ANNONA (ANNONACEAE)
sp. unspecified - *Arsenura armida* ssp., *Gonimbrasia hecate*.
muricata - *Attacus atlas*.

purpurea - *Automeris zugana, Periphoba arcaei.*
senegalensis - *Bunaeopsis hersilioides, Carnegia mirabilis, Imbrasia nictitans, Micragone ducorpsi, M. nenia, Pseudobunaea* sp.
squamosa - *Attacus caesar, Attacus lorquinii.*

ANOGEISSUS (COMBRETACEAE)
latifolia - *Antheraea paphia.*

APIUM (APIACEAE)
graveolens - *Samia cynthia* ssp.

APODYTES (ICACINACEAE)
dimidiata - *Lobobunaea angasana.*

ARBUTUS (ERICACEAE)
sp. unspecified - *Hyalophora euryalus euryalus.*
menziesii - *Antheraea polyphemus* ssp., *Citheronia splendens sinaloensis, Saturnia mendocino.*
unedo - *Saturnia pavonia* ssp., *S. walterorum.*

ARCTOSTAPHYLOS (ERICACEAE)
sp. unspecified - *Hemileuca eglanterina* ssp., *Hyalophora euryalus euryalus, H. columbia gloveri, Saturnia walterorum.*
alpinus - *Saturnia mendocino.*
glauca - *Hemileuca eglanterina* ssp., *Hyalophora euryalus euryalus, Saturnia mendocino.*
patula - *Hyalophora columbia gloveri, H. euryalus euryalus.*
pungens - *Citheronia splendens sinaloensis.*
tomentosa - *Saturnia mendocino.*
uva-ursi - *Hemileuca eglanterina* ssp., *Saturnia mendocino.*

ARDISIA (MYRSINACEAE)
revoluta - *Periphoba arcaei.*
sieboldii - *Attacus atlas.*

ARONIA (ROSACEAE)
arbutifolia - *Hyalophora cecropia.*

ARTEMISIA (ASTERACEAE)
sp. unspecified - *Hemileuca hera hera.*
californica - *Vegetia ducalis.*
cana - *Hemileuca hera hera.*
filifolia - *Hemileuca magnifica.*

tridentata - *Hemileuca hera hera, H. hera marcata, H. magnifica.*
vulgaris - *Attacus atlas.*

ARTOCARPUS (MORACEAE)
sp. unspecified - *Cricula trifenestrata.*
integer - *Attacus caesar, A. lorquinii.*

ASTRONIUM (ANACARDIACEAE)
graveolens - *Eacles imperialis* ssp., *Periphoba arcaei.*

AVERRHOA (OXALIDACEAE)
bilimbi - *Actias maenas.*
carambola - *Attacus atlas.*

AZADIRACHTA (MELIACEAE)
indica - *Actias selene* ssp., *Samia cynthia* ssp.

BACCHARIS (ASTERACEAE)
sculpta - *Rothschildia condor.*

BANISTERIOPSIS (MALPIGHIACEAE)
muricata - *Hylesia lineata.*

BAPTISIA (FABACEAE: FABOIDEAE)
tinctoria - *Antheraea polyphemus* ssp., *Automeris io* ssp., *A. io io.*

BASSIA (CHENOPODIACEAE)
latifolia - *Antheraea paphia, A. roylei.*
longifolia - *Antheraea paphia, A. roylei.*

BAUHINIA (FABACEAE: CAESALPINIOIDEAE)
sp. unspecified - *Gonimbrasia hecate, Holocerina angulata, H. smilax, Melanocera menippe, Pseudobunaea irius.*
forficata - *Automeris complicata.*
petersiana - *Imbrasia dione.*
reticulata - *Bunaea alcinoe, Holocerina agomensis, H. angulata, Imbrasia nictitans, Pseudobunaea pallens, Pseudobunaea* sp.
ungulata - *Hylesia lineata.*
variegata - *Antheraea paphia.*

BEQUAERTIODENDRON (SAPOTACEAE)
sp. unspecified - *Bunaeopsis hersilia.*

BERBERIS (BERBERIDACEAE)

sp. unspecified - *Callosamia promethea, Hyalophora cecropia, Samia cynthia* ssp.

asiatica - *Attacus atlas, A. taprobanis.*

flexuosa - *Leucanella stuarti* ssp.

lauina - *Hyperchiria incisa* ssp.

thunbergii - *Attacus atlas, A. taprobanis.*

vulgaris - *Attacus atlas, A. taprobanis, Callosamia promethea, Hyalophora cecropia.*

BERLINIA (FABACEAE: CAESALPINIOIDEAE)

sp. unspecified - *Gonimbrasia hecate, Gynanisa maja* ssp., *Imbrasia ertli, I. nictitans, Pseudobunaea epithyrena.*

paniculata - *Cinabra hyperbius, Goodia kuntzei, Gynanisa maja* ssp., *Micragone ansorgei, Pseudobunaea pallens.*

BETULA (BETULACEAE)

sp. unspecified - *Aglia tau* ssp., most *Antheraea* sp., *A. polyphemus* ssp., *Automeris io* ssp., *A. io io, Callosamia promethea, Eacles imperialis imperialis, Imbrasia krucki, Leucanella leucane, Opodiphthera helena, Saturnia pavonia* ssp., *S. pyri.*

alba - *Aglia tau* ssp., *Antheraea polyphemus* ssp., *Eacles imperialis* ssp., *Hyalophora cecropia.*

alba var. ***laciniata*** - *Eacles imperialis* ssp.

alleghaniensis - *Hyalophora cecropia.*

alnoides - *Antheraea knyvetti, A. roylei.*

excelsa - *Hyalophora cecropia.*

lenta - *Antheraea polyphemus* ssp., *Hyalophora cecropia.*

occidentalis - *Hyalophora euryalus euryalus.*

papyrifera - *Actias luna* ssp., *Antheraea polyphemus* ssp., *Caligula simla, Callosamia promethea, Eacles imperialis* ssp., *Eupackardia calleta, Hemileuca eglanterina* ssp., *Hyalophora cecropia.*

pendula - *Actias luna* ssp., *Aglia tau* ssp., *Antheraea polyphemus* ssp., *Hyalophora columbia gloveri, Saturnia lindia.*

platyphylla - *Actias artemis, Aglia tau* ssp., *Attacus atlas.*

populifolia - *Antheraea polyphemus* ssp., *Eacles imperialis* ssp., *Hyalophora cecropia, Hemileuca lucina.*

pumila - *Hemileuca maia.*

BISCHOFIA (EUPHORBIACEAE)

sp. unspecified - *Cricula trifenestrata.*

javanica - *Attacus atlas.*

BIXA (BIXACEAE)

sp. unspecified - *Citheronia phoronea.*

orellana - *Citheronia phoronea.*

BOMBACOPSIS (BOMBACACEAE)

quinatum - *Arsenura armida* ssp., *Caio championi, Hylesia lineata, Periphoba arcaei, Titaea tamerlan* ssp.

BOMBAX (BOMBACACEAE)

sp. unspecified - *Titaea tamerlan* ssp.

ceiba - *Arsenura armida* ssp., *Dysdaemonia boreas.*

heptaphyllum - *Antheraea paphia.*

malabaricum - *Antheraea paphia.*

BOUTELOUA (POACEAE)

curtipendula - *Hemileuca hualapai, H. oliviae.*

gracilis - *Hemileuca oliviae.*

hirsuta - *Hemileuca oliviae.*

oligostachya - *Hemileuca oliviae.*

BRACHYLAENA (ASTERACEAE)

discolor - *Ludia delegorguei.*

rotundata - *Ludia delegorguei.*

BRACHYOTUM (MELASTOMATACEAE)

ledifolium - *Copaxa semioculata orientalis.*

BRACHYSTEGIA (FABACEAE: CAESALPINIOIDEAE)

sp. unspecified - *Gynanisa maja* ssp., *Holocerina agomensis, Imbrasia carnegiei, I. nictitans, Pseudobunaea callista.*

globifera - *Athletes semialba, Imbrasia carnegiei.*

randii - *Athletes gigas, A. semialba, Cinabra hyperbius, Goodia kuntzei, Imbrasia ertli, I. macrothyris, Lobobunaea natalensis, Pseudobunaea epithyrena.*

spiciformis - *Athletes semialba, Cinabra hyperbius, Goodia kuntzei, Imbrasia ertli, I. macrothyris, Micragone ansorgei, Orthogonioptilum incana, Pseudobunaea irius.*

venosa - *Cinabra hyperbius, Goodia kuntzei, Gynanisa ata, G. maja* ssp., *Imbrasia macrops.*

BRASSAIA (ARALIACEAE)

actinophylla var. **arboricola** - *Attacus atlas.*

BRIDELIA (EUPHORBIACEAE)

micrantha - *Bunaeopsis aurantiaca.*

grandis - *Pseudobunaea cleopatra.*

BROMUS (POACEAE)
polyanthus - *Hemileuca oliviae.*

BROSIMUN (MORACEAE)
utile - *Automeris* sp.

BUCHLOE (POACEAE)
dactyloides - *Hemileuca oliviae.*

BURKEA (FABACEAE: CAESALPINIOIDEAE)
africana - *Cirina forda, Gonimbrasia belina, Rohaniella pygmaea.*

BURSERA (BURSERACEAE)
microphylla - *Hemileuca electra* ssp.
simaruba - *Citheronia lobesis* ssp., *Hylesia acuta.*
tomentosa - *Eacles imperialis* ssp., *E. imperialis decoris.*

BYRSONIMA (MALPIGHIACEAE)
crassifolia - *Hylesia lineata.*

BYTTNERIA (STERCULIACEAE)
aculeata - *Hylesia lineata.*
catalpaefolia - *Hylesia lineata.*

CALLIANDRA (FABACEAE: MIMOSOIDEAE)
emarginata - *Hylesia lineata.*

CALLUNA (ERICACEAE)
vulgaris - *Saturnia pavonia* ssp.

CALYCOPHYLLUM (RUBIACEAE)
candidissimum - *Automeris rubrescens* ssp., *Citheronia lobesis* ssp., *Hylesia lineata, Periphoba arcaei.*

CANANGA (ANNONACEAE)
odorata - *Attacus atlas, A. caesar, A. lorquinii.*

CANARIUM (BURSERACEAE)
sp. unspecified - *Actias maenas.*
album - *Cricula trifenestrata.*
commune - *Attacus atlas, Cricula trifenestrata.*

CANTHIUM (RUBIACEAE)
diccocum - *Antheraea paphia.*

CAPSICUM (SOLANACEAE)
annuum - *Antheraea polyphemus* ssp.

CAREYA (BARRINGTONIACEAE)
arborea - *Antheraea paphia, Cricula trifenestrata.*
sphaerica - *Antheraea paphia.*

CARICA (CARICACEAE)
papaya - *Samia cynthia* ssp.

CARISSA (APOCYNACEAE)
carandas - *Antheraea paphia.*
edulis - *Cirina forda.*
grandiflora - *Cirina forda, Gonimbrasia belina.*

CARPINUS (BETULACEAE)
sp. unspecified - *Actias luna* ssp., *Antheraea mylitta, Eacles imperialis imperialis, Periphoba hircia.*
americana - *Antheraea polyphemus* ssp., *Hyalophora cecropia.*
betulus - *Aglia tau* ssp., *Antheraea paphia, A. pernyi, Attacus atlas, Lonomia cynira, Samia cynthia* ssp.
caroliniana - *Anisota virginiensis, Antheraea polyphemus* ssp., *Automeris io* ssp., *Eacles imperialis* ssp.

CARYA (JUGLANDACEAE)
sp. unspecified - *Actias artemis, Actias luna* ssp., *Anisota senatoria, A. stigma* ssp., *Antheraea polyphemus* ssp., *A. yamamai* ssp., *Automeris io* ssp., *A. io io, Caligula simla, Citheronia brissotii* ssp., *C. mexicana, C. regalis, Eacles imperialis* ssp., *Hyalophora cecropia.*
amara - *Antheraea polyphemus* ssp.
cordiformis - *Actias luna* ssp.

glabra - *Actias luna* ssp., *Antheraea polyphemus* ssp., *Citheronia regalis, C. splendens sinaloensis.*
illinoensis - *Actias luna* ssp., *Citheronia brissotii* ssp., *C. regalis, C. splendens splendens, Hyalophora cecropia.*
ovata - *Actias luna* ssp., *Antheraea polyphemus* ssp., *Citheronia regalis.*
tomentosa - *Antheraea polyphemus* ssp.

CARYOCAR (CARYOCARACEAE)

brasiliensis - *Citheronia laocoon.*

CASEARIA (FLACOURTIACEAE)

arguta - *Hylesia dalina, H. lineata.*
corymbosa - *Hylesia lineata, Rothschildia lebeau* ssp.
odorata - *Hylesia dalina.*
sylvestris - *Hylesia lineata.*

CASSIA (FABACEAE: CAESALPINIOIDEAE)

sp. unspecified - *Athletes semialba, Gynanisa maja* ssp.
alata - *Periphoba arcaei.*
biflora - *Automeris io io, A. rubrescens* ssp., *Hylesia lineata, Periphoba arcaei.*
chamaecrista - *Automeris io io.*
corymbosa - *Hyperchiria incisa* ssp.
emarginata - *Syssphinx montana.*
fasciculata - *Automeris io* ssp.
fistula - *Samia cynthia* ssp.
grandis - *Automeris rubrescens* ssp., *Syssphinx molina.*
hayesina - *Automeris zugana.*
hookeriana - *Leucanella stuarti* ssp.
lanceolata - *Antheraea paphia.*

CASTANEA (FAGACEAE)

sp. unspecified - *Actias maenas, A. selene* ssp., *Anisota stigma fuscosa, Antheraea pernyi, A. polyphemus* ssp., *A. yamamai* ssp., *Caligula boisduvali* ssp., *Eacles imperialis* ssp., *Rhodinia fugax, Saturnia pyri.*
crenata - *Aglia tau* ssp., *Antheraea yamamai* ssp., *Caligula japonica arisana, C. japonica* ssp.
dentata - *Actias luna* ssp., *Anisota senatoria, A. stigma* ssp., *A. virginiensis, Antheraea polyphemus* ssp., *Eacles imperialis* ssp.
mollissima - *Actias artemis, A. selene ningpoana, Anisota stigma stigma, Caligula japonica* ssp.
pumila - *Eacles imperialis* ssp.
sativa - *Actias luna* ssp., *Antheraea polyphemus* ssp., *Eacles imperialis* ssp.
vesca - *Antheraea polyphemus* ssp.
vulgaris - *Anisota finlaysoni, A. peigleri, A. pellucida pellucida, A. senatoria, A. stigma fuscosa, A. stigma stigma.*

CATALPA (BIGNONIACEAE)

bungei - *Caligula japonica* ssp.

CEANOTHUS (RHAMNACEAE)

sp. unspecified - *Hemileuca eglanterina* ssp., *Hyalophora cecropia, H. euryalus euryalus.*

cordulatus - *Hyalophora columbia gloveri, H. euryalus euryalus, Saturnia albofasciata.*

cuneatus - *Saturnia albofasciata, S. mendocino.*

fendleri - *Automeris pamina.*

integerrimus - *Hyalophora euryalus euryalus.*

leucodermis - *Saturnia albofasciata.*

sanguineus - *Hyalophora euryalus euryalus.*

spinosa - *Hemileuca eglanterina* ssp., *Saturnia albofasciata.*

thyrsiflorus - *Hemileuca eglanterina* ssp., *Hyalophora euryalus euryalus.*

velutinus - *Hemileuca eglanterina* ssp., *Hyalophora columbia gloveri, H. euryalus euryalus.*

CECROPIA (MORACEAE)

sp. unspecified - *Automeris* sp.

CEDRELA (MELIACEAE)

odorata - *Dirphia avia, Eacles imperialis decoris.*

paniculata - *Actias selene* ssp.

CEDRUS (PINACEAE)

sp. unspecified - *Eacles imperialis* ssp.

CEIBA (BOMBACACEAE)

acuminata - *Caio richardsoni.*

pentandra - *Attacus atlas, Dysdaemonia boreas.*

CELASTRUS (CELASTRACEAE)

paniculatus - *Antheraea paphia.*

CELTIS (ULMACEAE)

sp. unspecified - *Agapema galbina* ssp., *Bunaea alcinoe, Eacles imperialis tucumana.*

kraussiana - *Pseudobunaea tyrrhena.*

laevigata - *Automeris io* ssp.

pallida - *Automeris io io, A. pamina, A. randa, Pseudobunaea tyrrhena, Rhodinia fugax.*

sinensis - *Rhodinia verecunda.*

spinosa - *Hyperchiria incisa* ssp.

tala - *Leucanella aspera* ssp.

CENTROSEMA (FABACEAE: FABOIDEAE)
pubescens - *Automeris zugana.*

CEPHALANTHUS (RUBIACEAE)
sp. unspecified - *Hyalophora cecropia, H. euryalus euryalus, Rothschildia aurota speculifera, R. orizaba* ssp.
occidentalis - *Callosamia promethea, Citheronia regalis, Hyalophora cecropia.*

CEPHALARIA (DIPSACACEAE)
procera - *Saturnia cephalariae, S. pyri.*

CERCIDIUM (FABACEAE: CAESALPINIOIDEAE)
sp. unspecified - *Hemileuca tricolor.*
floridum - *Syssphinx heiligbrodti.*
microphyllum - *Hemileuca tricolor, Syssphinx heiligbrodti.*

CERCIS (FABACEAE: CAESALPINIOIDEAE)
canadensis - *Automeris io* ssp., *A. io lilith, A. zephyria.*
siliquastrum - ?*Athletes gigas, Caligula japonica* ssp.

CERCOCARPUS (ROSACEAE)
sp. unspecified - *Hemileuca chinatiensis.*
betuloides - *Hemileuca eglanterina* ssp., *H. griffini, H. nuttalli nuttalli, Saturnia albofasciata.*
betuloides var. ***blancheri*** - *Automeris pamina, A. zephyria.*
montanus - *Hemileuca eglanterina* ssp., *Saturnia albofasciata.*
parvifolius - *Automeris pamina.*

CHAMAECYPARIS (CUPRESSACEAE)
thyoides - *Eacles imperialis* ssp.

CHILOPSIS (BIGNONIACEAE)
linearis - *Automeris iris hesselorum.*

CHLORIS (POACEAE)
virgata - *Decachorda aspersa.*

CHLOROPHORA (MORACEAE)
sp. unspecified - *Imbrasia dione.*

CHLOROXYLON (RUTACEAE)
swietenia - *Antheraea paphia.*

CHOMELIA (RUBIACEAE)

spinosa - *Hylesia lineata.*

CHORISIA (BOMBACACEAE)

sp. unspecified - *Caio championi, C. richardsoni, Dysdaemonia boreas.*
insignis - *Arsenura armida* ssp., *A. polyodonta.*
speciosa - *Arsenura armida* ssp., *Dysdaemonia fosteri.*

CHRYSOPHYLLUM (SAPOTACEAE)

sp. unspecified - *Attacus lorquinii, Bunaeopsis hersilia.*

CINCHONA (RUBIACEAE)

succirubra - *Attacus atlas, Cricula trifenestrata.*

CINNAMOMUM (LAURACEAE)

camphora - *Actias selene* ssp., *A. selene ningpoana, Antheraea frithi javanensis, A. pernyi, Attacus atlas, A. taprobanis, Caligula japonica arisana, C. japonica* ssp., *Callosamia promethea, C. securifera, Copaxa lavendera, Cricula trifenestrata, Eriogyna pyretorum, Samia cynthia walkeri.*
iners - *Attacus atlas, ?Cricula trifenestrata.*
obtusifolium - *Antheraea assamensis.*
zeylanicum - *Attacus atlas, A. taprobanis, ?Cricula trifenestrata.*

CIPADESSA (MELIACEAE)

fruticosa - *Antheraea paphia.*

CISSUS (VITACEAE)

sp. unspecified - *Loepa katinka* ssp., *L. sikkima.*

CITRUS (RUTACEAE)

sp. unspecified - *Attacus atlas, A. caesar, A. lorquinii, Automeris melanops, Pseudautomeris salmonea, Rothschildia lebeau forbesi, R. lebeau* ssp.
aurantium - *Automeris complicata.*
reticulata - *Loepa oberthuri.*

CLERODENDRUM (VERBENACEAE)

infortunatum - *Attacus atlas.*
serratum - *Attacus atlas.*

CLIDEMIA (MELASTOMATACEAE)

hirta - *Attacus atlas.*

COCHLOSPERMUM (COCHLOSPERMACEAE)
vitifolium - *Citheronia lobesis* ssp., *Eacles imperialis* ssp., *E. imperialis decoris.*

COFFEA (RUBIACEAE)
sp. unspecified - *Hyalophora euryalus euryalus.*
arabica - *Attacus atlas.*

COLEOGYNE (ROSACEAE)
ramosissima - *Hemileuca griffini.*

COLEUS (LAMIACEAE)
barbatus - *Ludia orinoptena.*
blumei - *Ludia orinoptena.*

COLOPHOSPERMUM (FABACEAE: CAESALPINIOIDEAE)
sp. unspecified - *Gynanisa maja* ssp.
mopane - *Gonimbrasia belina, Gynanisa nigra.*

COMBRETUM (COMBRETACEAE)
sp. unspecified - *Gonimbrasia hecate, Imbrasia nictitans, Pseudaphelia apollinaris.*

COMMIPHORA (BURSERACEAE)
caryaefolia - *Usta terpsichore.*
mollis - *Argema besanti, A. mimosae* ssp.

COMPTONIA (MYRICACEAE)
peregrina - *Automeris io* ssp.

CONDALIA (RHAMNACEAE)
sp. unspecified - *Hemileuca chinatiensis.*
ericoides - *Agapema galbina galbina, Hemileuca conwayae.*
hookeri - *Agapema galbina* ssp.
obovata - *Agapema solita.*
spathulata - *Agapema galbina anona.*

CONOCARPUS (COMBRETACEAE)
erectus - *Automeris io lilith.*

CONOPODIUM (APIACEAE)
denudatum - *Eudyaria venata.*

COPAIFERA (FABACEAE: CAESALPINIOIDEAE)
sp. unspecified - *Gonimbrasia belina, Nudaurelia belina.*

CORDIA (BORAGINACEAE)
alliodora - *Automeris* sp., *A. rubrescens* ssp., *Hylesia lineata*.

CORDYLA (FABACEAE: FABOIDEAE)
africana - ?*Antistathmoptera daltonae*.

CORIARIA (CORIARIACEAE)
nepalensis - *Actias selene* ssp., *Samia cynthia* ssp.

CORNUS (CORNACEAE)
sp. unspecified - *Antheraea polyphemus* ssp., *Automeris io io*, *Eacles imperialis* ssp., *Samia cynthia* ssp., *Saturnia pavonia* ssp.
controversa - *Caligula boisduvali* ssp.
florida - *Automeris io* ssp.
racemosa - *Antheraea polyphemus* ssp.
stolonifera - *Hyalophora cecropia*.

CORYLUS (BETULACEAE)
sp. unspecified - *Actias luna* ssp., *Anisota manitobensis*, *A. senatoria*, *A. stigma* ssp., *A. stigma stigma*, *A. virginiensis*, *Antheraea polyphemus* ssp., *Automeris io* ssp., *Callosamia promethea*, *Citheronia regalis*, *Hemileuca maia*, *H. nevadensis*, *Imbrasia rubra*, *Saturnia pyri*.
americana - *Antheraea polyphemus* ssp., *Hyalophora cecropia*.
avellana - *Aglia tau* ssp., *Automeris io* ssp., *Saturnia lindia*.
colurna - *Actias selene* ssp.
cornuta - *Hyalophora cecropia*.
heterophylla - *Aglia tau* ssp., *Caligula boisduvali fallax*, *C. japonica* ssp.
rostrata - *Antheraea polyphemus* ssp., *Hyalophora cecropia*.

COUTAREA (RUBIACEAE)
hexandra - *Rothschildia erycina* ssp.

CRATAEGUS (ROSACEAE)
sp. unspecified - *Actias selene* ssp., *Antherina suraka*, *Automeris illustris*, *A. io io*, *A. io* ssp., *A. metzli*, *Bunaea alcinoe*, *Caligula cachara*, *C. japonica* ssp., *C. simla*, *Copaxa syntheratoides*, *Cricula andrei*, *Eacles imperialis* ssp., *Eriogyna pyretorum*, *Gonimbrasia tyrrhea* ssp., *Hyalophora cecropia*, *Imbrasia cleoris*, *I. cytherea*, *Lobobunaea acetes*, *Loepa katinka* ssp., *Periphoba hircia*, *Rhodinia newara*, *Saturnia pavonia* ssp.
calpodendron - *Hyalophora cecropia*.
coccinea - *Antheraea polyphemus* ssp., *Hyalophora cecropia*.
crus-galli - *Antheraea polyphemus* ssp., *Hyalophora cecropia*.

oxyacantha - *Caligula simla, Dirphia avia, Hyalophora cecropia, Paradirphia eumedide, Saturnia spini.*
pedicellata - *Hyalophora cecropia.*
pinnatifida - *Antheraea pernyi.*
tomentosa - *Antheraea polyphemus* ssp., *Hyalophora cecropia.*

CRESCENTIA (BIGNONIACEAE)
alata - *Automeris io io, A. rubrescens* ssp.

CROTON (EUPHORBIACEAE)
sp. unspecified - *Bunaea alcinoe, Epiphora rectifascia, Rothschildia maurus.*
floribundus - *Procitheronia principalis.*
gratissimus - *Epiphora mythimnia.*
oligondrum - *Epiphora vacuna, Pseudobunaea cleopatra.*
sylvaticus - *Epiphora mythimnia.*

CRYPTOCARYA (LAURACEAE)
peumus - *Catocephala socialis, Polythysana cinerascens.*
rubra - *Polythysana apollina, P. cinerascens, P. rubrescens.*

CUPANIA (SAPINDACEAE)
guatemalensis - *Hylesia lineata.*

CUPRESSUS (CUPRESSACEAE)
sp. unspecified - *Automeris* sp., *Citheronia laocoon, Eacles imperialis imperialis.*
macrocarpa - *Imbrasia cytherea.*

CURCUMA (ZINGIBERACEAE)
longa - *Attacus atlas.*

CUSSONIA (ARALIACEAE)
spicata - *Bunaea alcinoe, Holocerina rhodesiensis.*

CYCLOBALANOPSIS (FAGACEAE)
glauca - *Antheraea roylei, Attacus lorquinii, Rhodinia fugax, R. verecunda.*

CYDISTA (BIGNONIACEAE)
heterophylla - *Automeris zugana.*

CYDONIA (ROSACEAE)
sp. unspecified - *Imbrasia cytherea, Saturnia pyri.*
oblonga - *Antheraea polyphemus* ssp.

CYNODON (POACEAE)
sp. unspecified - *Automeris colenon, Hemileuca hualapai.*

CYPERUS (CYPERACEAE)
esculentus - *Antheraea roylei.*

DALBERGIA (FABACEAE: FABOIDEAE)
retusa - ?*Hylesia lineata.*

DALEA (FABACEAE: FABOIDEAE)
californica - *Hemileuca burnsi.*
fremontii - *Hemileuca burnsi.*

DAPHNIPHYLLUM (DAPHNIPHYLLACEAE)
himalense - *Antheraea roylei.*

DELONIX (FABACEAE: CAESALPINIOIDEAE)
sp. unspecified - *Gonimbrasia zambesina.*

DIALIUM (FABACEAE: CAESALPINIOIDEAE)
guineense - *Eustera brachyura.*

DIERVILLA (CAPRIFOLIACEAE)
sp. unspecified - *Citheronia regalis.*

DILLENIA (DILLENIACEAE)
indica - *Attacus atlas.*
pentagyna - *Attacus atlas, Loepa katinka* ssp.

DIOSPYROS (EBENACEAE)
sp. unspecified - *Automeris io io, Callosamia promethea, Eacles imperialis* ssp., *Gonimbrasia belina.*
kaki - *Citheronia laocoon.*
lycioides - *Gonimbrasia tyrrhea* ssp.
mespiliformis - *Gonimbrasia zambesina* ssp.
virginiana - *Actias luna* ssp., *Citheronia regalis.*

DIPHYSA (FABACEAE: FABOIDEAE)
robinioides - *Hylesia lineata.*

DIPLORHYNCHUS (APOCYNACEAE)
mossambicensis - *Imbrasia dione, I. rectilineata.*

DISCOGLYPREMNA (EUPHORBIACEAE)
caloneura - *Epiphora boolana, E. lugardi.*

DISTICHLIS (POACEAE)
spicata - *Hemileuca oliviae.*

DODONAEA (SAPINDACEAE)
viscosa - *Antheraea paphia.*

DOLICHOS (FABACEAE: FABOIDEAE)
sp. unspecified - *Bunaeopsis bomfordi.*

DYSOXYLUM (MELIACEAE)
sp. unspecified - *Coscinocera hercules.*

EKEBERGIA (MELIACEAE)
sp. unspecified - *Cinabra hyperbius.*
meyeri - *Bunaea alcinoe, Pseudobunaea tyrrhena, Cirina forda, Holocerina smilax.*
ruepellii - *Bunaea alcinoe.*

ELAEAGNUS (ELAEAGNACEAE)
angustifolia - *Hyalophora columbia gloveri.*
argentea - *Hyalophora columbia nokomis.*

ELAEOCARPUS (ELAEOCARPACEAE)
obovatus - *Opodiphthera* sp.

ELEPHANTORRHIZA (FABACEAE: MIMOSOIDEAE)
burchelli - *Gynanisa maja* ssp.

ENTEROLOBIUM (FABACEAE: MIMOSOIDEAE)
cyclocarpum - *Hylesia lineata.*

EPHEDRA (EPHEDRACEAE)
sp. unspecified - *Hemileuca griffini.*
nevadensis var. **aspera** - *Hemileuca chinatiensis.*

ERICA (ERICACEAE)
sp. unspecified - *Saturnia pavonia* ssp.

ERIOBOTRYA (ROSACEAE)
japonica - *Eriogyna pyretorum.*

ERIOCEPHALUS (ASTERACEAE)

africanus - *Vegetia ducalis.*
ericoides - *Vegetia ducalis, V. dewitzi.*
racemosus - *Vegetia grimmia.*
umbellatus - *Vegetia ducalis.*

ERIOGONUM (POLYGONACEAE)

sp. unspecified - *Hemileuca hera hera.*
effusum - *Hemileuca magnifica.*
fasciculatum - *Hemileuca burnsi, H. electra* ssp., *H. neumoegeni.*
fasciculatum var. ***polifolium*** - ?*Hemileuca burnsi, H. electra* ssp.
wrightii - *Hemileuca lex.*

ERIOSEMA (FABACEAE: FABOIDEAE)

sp. unspecified - *Vegetia ducalis.*

ERYNGIUM (APIACEAE)

agavifolium - *Eudyaria venata.*

ERYTHRINA (FABACEAE: FABOIDEAE)

sp. unspecified - *Attacus atlas, A. caesar, Automeris janus.*
abyssinica - *Urota sinope.*
caffra - *Urota sinope.*
crista-galli - *Eacles imperialis cacicus, E. imperialis opaca, Leucanella viridescens* ssp.
rubra - *Automeris metzli.*
subumbrana - *Attacus atlas, A. caesar.*
tomentosa - *Urota sinope.*

ERYTHROPHLEUM (FABACEAE: CAESALPINIOIDEAE)

guineense - *Cirina forda, Pseudimbrasia deyrollei.*

ERYTHROXYLUM (ERYTHROXYLACEAE)

havanense - *Hylesia lineata, Imbrasia dione.*

EUCALYPTUS (MYRTACEAE)

sp. unspecified - *Argema besanti, A. mimosae* ssp., *Automeris incarnata, Citheronia vogleri, Lobobunaea angasana, Opodiphthera loranthi.*
baxteri - *Opodiphthera helena.*
blakelyi - *Opodiphthera eucalypti.*
botryoides - *Opodiphthera eucalypti.*
bridgesiana - *Opodiphthera eucalypti.*
cinerea - *Opodiphthera eucalypti.*

eugenioides - *Opodiphthera eucalypti.*
fastigiata - *Opodiphthera eucalypti.*
globulus - *Imbrasia cytherea, Opodiphthera eucalypti.*
gunnii - *Actias maenas, Cricula trifenestrata, Opodiphthera eucalypti.*
leucoxylon - *Opodiphthera eucalypti.*
leucoxylon var. ***macrocarpa*** - *Opodiphthera eucalypti.*
macarthuri - *Opodiphthera eucalypti.*
maculosa - *Opodiphthera eucalypti.*
macrorhyncha - *Opodiphthera eucalypti.*
maideni - *Opodiphthera eucalypti.*
melliodora - *Opodiphthera eucalypti.*
nitens - *Opodiphthera eucalypti.*
odorata - *Opodiphthera helena.*
obliqua - *Opodiphthera eucalypti.*
paniculata - *Opodiphthera eucalypti.*
pauciflora - *Opodiphthera eucalypti.*
regnans - *Opodiphthera eucalypti.*
saligna - *Opodiphthera eucalypti.*
viminalis - *Opodiphthera eucalypti.*

EUCLEA (EBENACEAE)
sp. unspecified - *Bunaeopsis oubie, B. zaddachi.*
daphnoides - *Imbrasia cytherea.*
polyandra - *Imbrasia cytherea.*
racemosa - *Imbrasia cytherea.*
undulata - *Cirina forda.*

EUGENIA (MYRTACEAE)
sp. unspecified - *Antherina suraka, Imbrasia dione, Lobobunaea saturnus.*
aquea - *Cricula trifenestrata.*
cordata - *Lobobunaea angasana.*
cuminii - *Antheraea paphia.*
cuneifolia - *Argema mittrei.*
jambolana - *Antheraea paphia.*
malaccense - *Cricula trifenestrata.*
salamensis - *Periphoba arcaei.*

EUODIA (RUTACEAE)
fraxinifolia - *Antheraea roylei, Samia cynthia* ssp.

EUPATORIUM (ASTERACEAE)
buniifolium - *Rothschildia tucumana.*

EUPHORBIA (SAPINDACEAE)
longana - *Attacus atlas.*

EUSCAPHIS (STAPHYLEACEAE)
japonica - *Samia cynthia walkeri.*

EXCOECARIA (EUPHORBIACEAE)
biglandulosa - *Citheronia brissotii* ssp.

EXOSTEMA (RUBIACEAE)
mexicanum - *Rothschildia erycina* ssp., *R. lebeau* ssp.

EYSENHARDTIA (FABACEAE: FABOIDEAE)
polystachya - *Automeris iris hesselorum.*

FAGARA (RUTACEAE)
heitzii - *Epiphora vacuna.*
macrophylla - *Epiphora albida, E. ploetzi.*

FAGUS (FAGACEAE)
sp. unspecified - *Actias luna* ssp., *Aglia tau* ssp., *Anisota senatoria, Antheraea mylitta, A. pernyi, A. polyphemus* ssp., *A. yamamai* ssp., *Antherina suraka, Automerella flexuosa, Automeris beckeri, A. hamata, A. illustris, A. incarnata, A. io* ssp., *A. io io, A. io coloradensis, A. janus, A. jucunda, Bunaeopsis oubie, Callosamia promethea, Dirphia curitiba, Eacles ducalis, E. imperialis* ssp., *E. imperialis imperialis, Gonimbrasia zambesina* ssp., *Hemileuca lex, Hyperchiria incisa* ssp., *H. nausica, Imbrasia krucki, I. oyemensis, Leucanella aspera, Lonomia cynira, Periphoba hircia, Pseudantheraea discrepans, Pseudobunaea alinda, P. pallens, P. tyrrhena, Rhodinia fugax, R. jankowskii, Saturnia pavonia* ssp.
crenata - *Actias artemis, Aglia tau* ssp., *Caligula boisduvali* ssp.
ferruginea - *Antheraea polyphemus* ssp., *Hyalophora cecropia.*
grandifolia - *Dryocampa rubicunda* ssp., *Eacles imperialis* ssp.
longipetiolata - *Aglia tau* ssp.
pumilio - *Adetomeris erythrops.*
sylvatica - *Aglia tau* ssp., *Dirphia avia, D. baroma, Eacles imperialis* ssp., *E. penelope, Pseudodirphia eumedide.*

FAIDHERBIA (FABACEAE: MIMOSOIDEAE)
albida - *Cirina forda.*

FALLUGIA (ROSACEAE)
paradoxa - *Hemileuca neumoegeni.*

FICUS (MORACEAE)

sp. unspecified - ***Attacus atlas, ?A. erebus, Gonimbrasia belina, Goodia lunata.***
benjamina - ***Antheraea paphia, Automeris illustris, Hyperchiria incisa*** ssp.
cordata - ***Melanocera menippe.***
religiosa - ***Antheraea paphia.***
retusa - ***Antheraea paphia.***

FIRMIANA (STERCULIACEAE)

simplex - *Samia cynthia* ssp.

FORESTIERA (OLEACEAE)

angustifolia - *Eupackardia calleta, Hemileuca chinatiensis.*

FORSYTHIA (OLEACEAE)

sp. unspecified - *Samia cynthia* ssp.

FOUQUIERIA (FOUQUIERIACEAE)

splendens - *Eupackardia calleta.*

FRAGARIA (ROSACEAE)

sp. unspecified - *Saturnia pavonia* ssp.

FRAXINUS (OLEACEAE)

sp. unspecified - *Adetomeris erythrops, Antheraea polyphemus* ssp., *Attacus atlas, A. lorquinii, Automeris io* ssp., *A. io io, Callosamia promethea, Citheronia regalis, Eupackardia calleta, Hyalophora cecropia, Rothschildia aurota speculifera, R. jorulla* ssp., *R. lebeau forbesi, R. lebeau* ssp., *Samia cynthia* ssp., *Saturnia pavonia* ssp.
americana - *Callosamia promethea, Rothschildia orizaba* ssp.
berlandierana - *Saturnia atlantica.*
excelsior - *Saturnia pyri.*
pennsylvanica - *Callosamia promethea.*

FREMONTIA (STERCULIACEAE)

californica - *Hemileuca chinatiensis, Saturnia albofasciata.*

GALACTIA (FABACEAE: FABOIDEAE)

sp. unspecified - *Automeris io lilith.*

GAYLUSSACIA (ERICACEAE)

dumosa - *Antheraea polyphemus* ssp.
frondosa - *Hyalophora cecropia.*

GEIJERA (RUTACEAE)
salicifolia - *Opodiphthera eucalypti.*

GEOFFROEA (FABACEAE: FABOIDEAE)
decorticans - *Automeris amanda* ssp.

GINKGO (GINKGOACEAE)
biloba - *Caligula japonica arisana.*

GLEDITSIA (FABACEAE: CAESALPINIOIDEAE)
sp. unspecified - *Antheraea polyphemus* ssp., *Automeris io draudtiana.*
texana - *Syssphinx raspa.*
triacanthos - *Antheraea polyphemus* ssp., *Automeris banus* ssp., *A. godartii, A. io io, A. io* ssp., *Cicia crocata, Eacles imperialis* ssp., *E. imperialis imperialis, Hemileuca burnsi, H. juno, H. tricolor, Hyalophora cecropia, Othorene hodeva, Psilopygida crispula, Syssphinx bicolor, S. bisecta, S. blanchardi, S. heiligbrodti, S. hubbardi, S. molina, S. montana, S. quadrilineata.*

GLIRICIDIA (FABACEAE: FABOIDEAE)
sepium - *Automeris io io, ?Hylesia lineata.*

GLOCHIDION (EUPHORBIACEAE)
sp. unspecified - *Coscinocera hercules.*
hongkongense - *Caligula thibeta okurai, Samia cynthia walkeri.*
rubrum - *Attacus atlas.*
velutinum - *Attacus atlas.*

GLYCYRRHIZA (FABACEAE: FABOIDEAE)
sp. unspecified - *Automeris io coloradensis, A. io neomexicana.*

GMELINA (VERBENACEAE)
arborea - *Samia cynthia* ssp.

GORDONIA (THEACEAE)
lasianthus - *Callosamia promethea, C. securifera.*

GOSSYPIUM (MALVACEAE)
sp. unspecified - *Citheronia beledonon.*
herbaceum - *Automeris io* ssp., *A. io io, Citheronia laocoon, C. regalis, Eacles imperialis* ssp.
hirsutum - *Automeris io lilith.*
thurberi - *Citheronia splendens sinaloensis, C. splendens splendens.*

GOUANIA (RHAMNACEAE)
polygama - *Hylesia lineata.*

GRUMILEA (RUBIACEAE)
capensis - *Pseudobunaea tyrrhena.*

GUAZUMA (STERCULIACEAE)
ulmifolia - *Automeris rubrescens* ssp., *Hylesia lineata, Periphoba arcaei.*

GUETTARDA (RUBIACEAE)
macrosperma - *Hylesia lineata.*

GYMNOCLADUS (FABACEAE: CAESALPINIOIDEAE)
canadensis - *Syssphinx bicolor, S. bisecta.*
dioica - *Syssphinx bicolor.*

GYMNOSPORIA (CELASTRACEAE)
senegalensis - *Bunaea alcinoe.*

HAEMATOXYLUM (FABACEAE: CAESALPINIOIDEAE)
brasalita - *Syssphinx montana.*

HALESIA (STYRACACEAE)
carolina - *Callosamia promethea.*

HAMAMELIS (HAMAMELIDACEAE)
virginiana - *Antheraea polyphemus* ssp.

HARPEPHYLLUM (ANACARDIACEAE)
caffrum - *Bunaea alcinoe.*

HELICONIA (MUSACEAE)
sp. unspecified - *Attacus lorquinii.*

HELINUS (RHAMNACEAE)
ovatus - *Epiphora mythimnia.*

HETEROPANAX (ARALIACEAE)
fragrans - *Samia cynthia* ssp.

HETEROPOGON (POACEAE)
sp. unspecified - *Bunaeopsis zaddachi.*

HEVEA (EUPHORBIACEAE)
brasiliensis - *Rhodinia fugax*.

HIBISCUS (MALVACEAE)
sp. unspecified - *Actias selene* ssp., *Attacus atlas*, *Automeris io lilith*.
lasiocarpos - *Automeris io* ssp.
syriacus - *Actias selene ningpoana*.
tiliaceus - *Automeris melanops*.

HILARIA (POACEAE)
cenchroides - *Hemileuca oliviae*.

HIPPOBROMUS (SAPINDACEAE)
pauciflora - *Cirina forda*.

HIPPOPHAE (ELAEAGNACEAE)
rhamnoides - *Saturnia pavonia* ssp.

HIRTELLA (CHRYSOBALANACEAE)
racemosa - *Hylesia lineata*.

HOLARRHENA (APOCYNACEAE)
floribunda - *Imbrasia nictitans*.

HOLODISCUS (ROSACEAE)
discolor - *Hemileuca eglanterina* ssp.

HOMALANTHUS (EUPHORBIACEAE)
populifolia - *Coscinocera hercules*.

HORDEUM (POACEAE)
caespitosum - *Hemileuca oliviae*.

HUMULUS (MORACEAE)
sp. unspecified - *Saturnia pavonia* ssp., *S. pyri*.
lupulus - *Actias luna* ssp., *Automeris io* ssp., *Hyalophora cecropia*.

HYMENAEA (FABACEAE: CAESALPINIOIDEAE)
altissima - *Schausiella arpi, Syssphinx bicolor*.
courbaril - *Automeris zugana, Dirphia avia, Hylesia lineata, Periphoba arcaei, Schausiella santarosensis*.

HYPARRHENIA (POACEAE)
sp. unspecified - *Decachorda rosea.*
contracta - *Bunaeopsis hersilia.*

HYPHAENE (ARECACEAE)
crinita - *Imbrasia anna.*

HYPTIS (LAMIACEAE)
pectinata - *Hylesia lineata.*

IDRIA (FOUQUIERIACEAE)
columnaris - *Hemileuca electra* ssp., *H. tricolor.*

ILEX (AQUIFOLIACEAE)
sp. unspecified - *Samia cynthia* ssp.
chinensis - *Archaeoattacus edwardsii, Attacus atlas, Samia cynthia* ssp.
paraguaiensis - *Citheronia laocoon.*
verticillata - *Actias artemis, Aglia tau* ssp., *Automeris io* ssp., *Rhodinia fugax.*

INGA (FABACEAE: MIMOSOIDEAE)
sp. unspecified - *Adeloneivaia fallax.*
vera - *Automeris rubrescens* ssp., *A. zugana, ?Hylesia lineata, Periphoba arcaei.*

IPOMOEA (CONVOLVULACEAE)
batatas - *Attacus atlas.*

IRIS (IRIDACEAE)
sp. unspecified - *Pseudautomeris ophthalmica.*

IRVINGIA (SIMAROUBACEAE)
sp. unspecified - *Bunaeopsis hersilia.*

JASMINUM (OLEACEAE)
pubigerum - *Holocerina smilax.*

JATROPHA (EUPHORBIACEAE)
sp. unspecified - *Rothschildia cincta* ssp.
curcas - *Imbrasia dione, Samia cynthia* ssp.
eurcas - *Rothschildia jorulla* ssp.

JOANNESIA (EUPHORBIACEAE)
princeps - *Citheronia laocoon.*

JUGLANS (JUGLANDACEAE)

sp. unspecified - *Actias truncatipennis, Citheronia brissotii* ssp., *C. mexicana, C. splendens splendens, Eacles ducalis, Eacles imperialis oslari, Saturnia pavonia* ssp.

cathayensis - *Caligula japonica arisana.*

cinerea - *Actias luna* ssp., *Antheraea polyphemus* ssp., *Citheronia regalis, Dryocampa rubicunda* ssp., *Eacles imperialis* ssp., *Samia cynthia* ssp.

neotropica - *Automeris* sp.

nigra - *Actias luna* ssp., *Antheraea polyphemus* ssp., *Citheronia regalis, C. splendens sinaloensis.*

regia - *Actias artemis, A. luna* ssp., *A. selene* ssp., *Aglia tau* ssp., *Argema besanti, A. mimosae* ssp., *Automeris io io, Caligula cachara, C. japonica* ssp., *C. simla, Citheronia regalis, C. splendens sinaloensis, Eacles imperialis imperialis, E. penelope, Hyalophora cecropia, Rhodinia newara, Samia cynthia* ssp., *Saturnia lindia, S. pyri.*

JULBERNARDIA (FABACEAE: CAESALPINIOIDEAE)

sp. unspecified - *Gynanisa ata, G. maja* ssp., *Imbrasia macrothyris, Pseudobunaea callista.*

paniculata - *Athletes gigas, Cinabra hyperbius, Goodia kuntzei, Imbrasia ertli, Micragone ansorgei, ?Pseudobunaea irius.*

JUNIPERUS (CUPRESSACEAE)

sp. unspecified - *Eacles imperialis imperialis.*

communis - *Eacles imperialis* ssp.

virginiana - *Eacles imperialis* ssp.

KALMIA (ERICACEAE)

sp. unspecified - *Callosamia promethea.*

angustifolia - *?Hyalophora columbia columbia.*

latifolia - *Attacus atlas.*

KHAYA (MELIACEAE)

sp. unspecified - *Imbrasia dione.*

nyasica - *?Gynanisa ata.*

senegalensis - *Bunaea alcinoe.*

KOELREUTERIA (SAPINDACEAE)

paniculata - *Eacles imperialis* ssp.

LABURNUM (FABACEAE: FABOIDEAE)

sp. unspecified - *Automerella flexuosa, Automeris boucardi, A. hamata, A. incarnata, Eacles penelope, Gynanisa maja* ssp., *Holocerina agomensis, Hyperchiria incisa*

ssp., *H. nausica, Imbrasia ertli, I. oyemensis, Leucanella aspera* ssp., *Urota sinope.*
anagyroides - *Leucanella memusae gardineri, Pseudodirphia eumedide.*

LACTUCA (ASTERACEAE)
sp. unspecified - *Ludia orinoptena.*

LAGERSTROEMIA (LYTHRACEAE)
indica - *Antheraea paphia, Attacus atlas, Automeris complicata, Caligula japonica* ssp., *Samia cynthia* ssp.
parviflora - *Antheraea paphia.*
subcostata - *Actias selene ningpoana.*

LANNEA (ANACARDIACEAE)
coromandelica - *Actias selene* ssp.
grandis - *Attacus atlas.*

LANTANA (VERBENACEAE)
camara - *Attacus atlas, Automeris zugana, Hylesia lineata.*

LARIX (PINACEAE)
sp. unspecified - *Hyalophora cecropia.*
decidua - *Hyalophora columbia gloveri.*
laricina - *Eacles imperialis* ssp., *Hyalophora columbia columbia, H. columbia gloveri, H. euryalus euryalus.*

LAURUS (LAURACEAE)
obtusifolia - *Antheraea assamensis.*

LAWSONIA (LYTHRACEAE)
alba - *Actias selene* ssp., *Samia cynthia* ssp.

LEEA (LEEACEAE)
sp. unspecified - *Loepa katinka* ssp., *L. sikkima.*

LESPEDEZA (FABACEAE: FABOIDEAE)
Lespedeza - *Automeris io* ssp.

LEUCAENA (FABACEAE: MIMOSOIDEAE)
leucocephala - *Automeris* sp.
pulverulenta - *Automeris io* ssp.

LEUCOPHYLLUM (SCROPHULARIACEAE)

frutescens - *Eupackardia calleta.*

LEUCOSPERMUM (PROTEACEAE)

ellipticum - *Imbrasia cytherea.*

LIGUSTRUM (OLEACEAE)

sp. unspecified - *Actias selene* ssp., ?*Agapema galbina* ssp., *Archaeoattacus edwardsii, Attacus atlas, A. lorquinii, A. taprobanis, Automeris coresus, A. illustris,* ?*A. io io, A. leucane, A. naranja* ssp., *A. tridens, Bunaea alcinoe, Callosamia promethea, Citheronia brissotii* ssp., *C. beledonon, C. hamifera* ssp., *C. regalis, Coscinocera hercules, Cricula andrei, Eacles ducalis, E. imperialis* ssp., *E. imperialis opaca, E. penelope, E. imperialis tucumana, Eupackardia calleta, Holocerina smilax, Hyalophora columbia gloveri, Imbrasia cleoris, Leucanella aspera, L. leucane, L. memusae* ssp., *L. viridescens* ssp., *Rhescyntis hippodamia* ssp., *Rothschildia arethusa* ssp., *R. aurota speculifera, R. aurota* ssp., *R. cincta* ssp., *R. erycina* ssp., *R. hesperus* ssp., *R. jacobaeae, R. jorulla* ssp., *R. lebeau forbesi, R. lebeau* ssp., *R. orizaba* ssp., *R. roxana, Samia cynthia* ssp., *Saturnia pyri,* ?*Syssphinx bicolor.*

japonicum - *Antherina suraka, Leucanella viridescens viridior, Rothschildia orizaba.*

ovalifolium - *Caligula thibeta extensa, Leucanella memusae gardineri.*

LINDERA (LAURACEAE)

benzoin - *Callosamia promethea, Eacles imperialis* ssp., *Samia cynthia* ssp.

megaphylla - *Samia cynthia walkeri.*

LIQUIDAMBAR (HAMAMELIDACEAE)

sp. unspecified - *Actias truncatipennis, Antherina suraka, Argema mimosae* ssp., *Lemaireia luteopeplus aureopeplus, Opodiphthera eucalypti, Polythysana apollina.*

formosana - *Actias selene ningpoana, A. sinensis heterogyna, Eriogyna pyretorum.*

styraciflua - *Actias luna* ssp., *A. maenas, A. selene* ssp., *A. sinensis heterogyna, Argema besanti, Automeris io* ssp., *A. io io, Callosamia promethea, Citheronia beledonon, C. regalis, C. splendens sinaloensis, Eacles imperialis oslari, E. imperialis imperialis, E. imperialis* ssp., *Hyalophora euryalus euryalus, Samia cynthia* ssp.

LIRIODENDRON (MAGNOLIACEAE)

sp. unspecified - *Automeris io* ssp.

tulipifera - *Antheraea polyphemus* ssp., *Automeris cecrops cecrops, A. io* ssp., *A. io io, Callosamia angulifera, C. promethea, C. securifera, Hyalophora cecropia, Imbrasia cytherea, Samia cynthia* ssp.

LITCHI (SAPINDACEAE)

chinensis - *Automeris io lilith.*

LITHREA (ANACARDIACEAE)

brasiliensis - *Citheronia splendens* ssp., *C. vogleri.*
molleoides - *Citheronia vogleri.*

LITSEA (LAURACEAE)

citrata - *Antheraea assamensis.*
polyantha - *Antheraea assamensis.*
salicifolia - *Antheraea assamensis.*

LONCHOCARPUS (FABACEAE: FABOIDEAE)

costaricensis - *Automeris zugana, Hylesia lineata.*
eriocarinalis - *Automeris zugana.*
minimiflorus - *Automeris rubrescens* ssp., *Hylesia lineata.*

LONICERA (CAPRIFOLIACEAE)

japonica - *Attacus atlas, Caligula boisduvali* ssp.

LORANTHUS (LORANTHACEAE)

acutifolius - *Citheronia vogleri.*
eugenioides - *Citheronia vogleri.*
flexicaulis - *Citheronia splendens* ssp.

LUEHEA (TILIACEAE)

speciosa - *Hylesia lineata.*

LUPINUS (FABACEAE: FABOIDEAE)

sp. unspecified - *Hemileuca hera hera.*

LYCIUM (SOLANACEAE)

sp. unspecified - *Agapema galbina anona.*

LYCURUS (POACEAE)

phleoides - *Hemileuca oliviae.*

LYSILOMA (FABACEAE: MIMOSOIDEAE)

auritum - *Adeloneivaia isara, Hylesia lineata, Periphoba arcaei.*
divaricata - *Adeloneivaia isara.*

LYTHRUM (LYTHRACEAE)

salicaria - *Saturnia pavonia* ssp.

MACHAERIUM (FABACEAE: FABOIDEAE)
brasiliensis - *Adelowalkeria flavosignata.*
kegelii - *Hylesia lineata.*

MACHILUS (LAURACEAE)
bombycina - *Antheraea assamensis.*
odoratissima - *Antheraea assamensis, Cricula trifenestrata.*

MACLURA (MORACEAE)
pomiferum - *Hyalophora cecropia.*

MAERUA (CAPPARIDACEAE)
sp. unspecified - *Imbrasia wahlbergi.*

MAESA (MYRSINACEAE)
sp. unspecified - *Antherina suraka, Bunaea alcinoe.*
alnifolia - *Imbrasia gueinzii.*
lanceolata - *Bunaea alcinoe, Nudaurelia bamendana.*

MAESOPSIS (RHAMNACEAE)
emini - *Epiphora rectifascia, E. vacuna.*

MAGNOLIA (MAGNOLIACEAE)
sp. unspecified - *Callosamia promethea.*
acuminata - *Callosamia promethea.*
grandiflora - *Callosamia securifera.*
sphenocarpa - *Antheraea assamensis.*
tripetala - *Callosamia angulifera.*
virginiana - *Attacus lorquinii, Automeris io* ssp., *Callosamia promethea, C. securifera.*

MALUS (ROSACEAE)
sp. unspecified - *Dirphia avia, Paradirphia boudinoti, Saturnia lindia.*
asiatica - *Actias selene ningpoana.*
coronaria - *Hyalophora cecropia.*
pumila - *Actias selene ningpoana, Antheraea polyphemus* ssp., *A. yamamai* ssp., *Automeris io* ssp., *Callosamia promethea, Hyalophora cecropia, H. euryalus euryalus.*

MALVAVISCUS (MALVACEAE)
arboreus - *Hylesia dalina, H. lineata.*
arboreus var. **drummondii** - *Automeris io* ssp.

MANGIFERA (ANACARDIACEAE)

sp. unspecified - *Actias selene* ssp., *Gonimbrasia zambesina* ssp., *Lobobunaea acetes*, *Nudaurelia anthina*, *Pseudantheraea arnobia*.

indica - *Attacus atlas*, *A. lorquinii*, *Citheronia splendens* ssp., *Cricula trifenestrata*, *Eacles imperialis cacicus*, *Imbrasia wahlbergi*.

MANGIUM (RHIZOPHORACEAE)

caseolari - *Antheraea paphia*.

MANIHOT (EUPHORBIACEAE)

esculenta - *Samia cynthia ricini*.

utilissima - *Samia cynthia* ssp.

MANILKARA (SAPOTACEAE)

chicle - *Copiopteryx semiramis* ssp., *Othorene purpurascens*.

MAYTENUS (CELASTRACEAE)

boaria - *Adetomeris erythrops*, *Polythysana apollina*, *P. cinerascens*, *P. rubrescens*.

MEDICAGO (FABACEAE: FABOIDEAE)

sp. unspecified - *Nudaurelia belina*.

MELASTOMA (MELASTOMATACEAE)

malabathricum - *Attacus atlas*.

MELIA (MELIACEAE)

azedarach - *Usta terpsichore*.

MELIANTHUS (MELIANTHACEAE)

dregianus - *Eochroa trimeni*.

minor - *Eochroa trimeni*.

pectinatus - *Eochroa trimeni*.

MELILOTUS (FABACEAE: FABOIDEAE)

sp. unspecified - *Automeris io* ssp.

alba - *Automeris io* ssp.

officinalis - *Automeris io* ssp.

MICHELIA (MAGNOLIACEAE)

sp. unspecified - *Antheraea assamensis*.

figo - *Samia cynthia* ssp.

MICONIA (MELASTOMATACEAE)
argentea - *Periphoba arcaei.*

MICROGLOSSA (ASTERACEAE)
mespilifolia - *Ludia delegorguei.*

MIKANIA (ASTERACEAE)
sp. unspecified - *Ludia orinoptena.*

MILNEA (MELIACEAE)
roxburghiana - *Attacus taprobanis.*

MIMOSA (FABACEAE: MIMOSOIDEAE)
sp. unspecified - *Adeloneivaia subangulata* ssp., *Adelowalkeria flavosignata, Argema mittrei, Automeris iris hesselorum, A. io* ssp., *Hemileuca tricolor, Imbrasia dione, Oiticella brevis, Rothschildia jorulla* ssp., *Syssphinx molina.*
borealis - *Hemileuca chinatiensis.*
dealbata - *Hemileuca rubridorsa.*
laurentzii - *Eubergia caisa.*
lindheimeri - *Hemileuca chinatiensis.*
pigra - *Automeris io io, Hylesia lineata.*
richelmanni - *Imbrasia carnegiei.*

MONOTES (DIPTEROCARPACEAE)
glaberi - *Holocerina agomensis.*
katangensis - *Argema kuhnei.*

MORINGA (MORINGACEAE)
oleifera - *Actias selene* ssp.

MORUS (MORACEAE)
sp. unspecified - *Antheraea yamamai* ssp., *Attacus atlas, Automeris io io.*
alba - *Hyalophora cecropia.*

MUEHLENBECKIA (POLYGONACEAE)
sagittaefolia - *Ormiscodes cinnamomea.*

MUHLENBERGIA (POACEAE)
texana - ?*Hemileuca oliviae.*

MUNTINGIA (ELAEOCARPACEAE)
calabura - *Hylesia lineata.*

MYRICA (MYRICACEAE)

sp. unspecified - *Imbrasia gueinzii.*
carolinensis - *Automeris io io, Hyalophora cecropia.*
cerifera - *Antheraea polyphemus* ssp., *Attacus caesar, A. lorquinii, Eacles imperialis* ssp., *Hyalophora cecropia.*
conifera - *Imbrasia cytherea.*
cordifolia - *Imbrasia cytherea.*
gale - *Saturnia pavonia* ssp.
pensylvanica - *Automeris io* ssp., *Callosamia promethea, Hyalophora columbia* ssp.

MYROSPERMUM (FABACEAE: FABOIDEAE)

frutescens - *Hylesia lineata.*

MYRSINE (MYRSINACEAE)

melanophloeos - *Imbrasia cytherea.*

NAUCLEA (RUBIACEAE)

rotundifolia - *Attacus atlas.*

NEGUNDO (ACERACEAE)

aceroides - *Hyalophora cecropia.*

NEMOPANTHUS (AQUIFOLIACEAE)

mucronata - ?*Hyalophora columbia columbia.*

NEPHELIUM (SAPINDACEAE)

litchi - *Attacus atlas, Automeris io* ssp.
lappaceum - *Attacus atlas.*

NERIUM (APOCYNACEAE)

oleander - *Antherina suraka, Attacus atlas, Gonimbrasia zambesina* ssp.

NICOLAIA (ZINGIBERACEAE)

speciosa - *Attacus atlas.*

NYSSA (NYSSACEAE)

sylvatica - *Actias luna* ssp., *Citheronia regalis.*

OCHNA (OCHNACEAE)

atropurpurea - *Melanocera menippe.*
schweinfurtiana - *Melanocera parva.*

OCIMUM (LAMIACEAE)

sp. unspecified - ?*Attacus taprobanis, Ludia orinoptena.*

OCOTEA (LAURACEAE)

veraguensis - *Copaxa moinieri.*

OLEA (OLEACEAE)

europaea - *Saturnia atlantica, S. pyri.*
paniculata - *Opodiphthera eucalypti.*

OPUNTIA (CACTACEAE)

sp. unspecified - *Automeris pamina.*

OSTRYA (BETULACEAE)

virginiana - *Actias luna* ssp., *Antheraea polyphemus* ssp., *Automeris io* ssp., *Eacles imperialis* ssp., *Hyalophora cecropia.*

OURATEA (OCHNACEAE)

lucens - *Hylesia lineata.*

OXYDENDRUM (ERICACEAE)

arboreum - *Citheronia regalis.*

PACHYCORMUS (ANACARDIACEAE)

discolor - *Hyalophora euryalus cedrosensis.*

PAEONIA (PAEONIACEAE)

officinalis - *Hyalophora cecropia.*

PAEPALANTHUS (ERIOCAULACEAE)

polyanthus - *Ithomisa* sp.

PANAX (ARALIACEAE)

elegans - *Coscinocera hercules.*

PAPPEA (SAPINDACEAE)

sp. unspecified - *Micragone cana.*
capensis - *Cirina forda.*

PARKIA (FABACEAE: MIMOSOIDEAE)

biglobosa - *Aurivillius arata.*
intermedia - *Attacus atlas.*
speciosa - *Attacus atlas.*

PARKINSONIA (FABACEAE: CAESALPINIOIDEAE)

microphyllum - *Syssphinx hubbardi, S. quadrilineata* ssp.
torreyana - *Syssphinx hubbardi.*

PARTHENOCISSUS (VITACEAE)

quinquefolia - *Hyalophora cecropia.*
tricuspidata - *Loepa katinka* ssp.

PASSIFLORA (PASSIFLORACEAE)

sp. unspecified - *Automeris complicata.*
quadrangulare - ?*Pseudautomeris irene* ssp., *P. salmonea.*

PAULLINIA (SAPINDACEAE)

sp. unspecified - *Epiphora albida.*
cururu - *Hylesia lineata.*

PELTOPHORUM (FABACEAE: CAESALPINIOIDEAE)

sp. unspecified - *Gynanisa maja* ssp.
africanum - *Aurivillius arata.*

PENNISETUM (POACEAE)

Pennisetum - *Bunaeopsis hersilia.*

PENTAPTERA (COMBRETACEAE)

glabra - *Antheraea paphia.*
tomentosa - *Antheraea paphia.*

PERSEA (LAURACEAE)

sp. unspecified - *Antheraea assamensis, Copaxa andensis C. denda, C. escalantei, C. joinvillea, C. lavendera.*
americana - *Attacus atlas, A. caesar, A. lorquinii, Automeris io* ssp., *A. io lilith, Copaxa andensis, C. decrescens, C. denda, C. escalantei, C. lavendera, C. mazaorum, C. multifenestrata multifenestrata, C. multifenestrata* ssp., *C. rufinans, Cricula trifenestrata, Hylesia alinda, ?H. lineata, Polythysana apollina.*
borbonia - *Attacus lorquinii.*
gratissima - *Attacus atlas, A. caesar, A. lorquinii.*

PHASEOLUS (FABACEAE: FABOIDEAE)

vulgaris - *Automeris io io.*

PHELLODENDRON (RUTACEAE)

amurense - *Rhodinia fugax, R. jankowskii.*

PHILADELPHUS (SAXIFRAGACEAE)
inodorus - *Hyalophora cecropia.*

PHOEBE (LAURACEAE)
porphyria - *Copaxa canella.*

PHORADENDRON (LORANTHACEAE)
quadrangulare - *Citheronia bellavista* ssp., *C. lobesis.*

PHOTINIA (ROSACEAE)
arbutifolia - *Hyalophora euryalus euryalus.*

PHRYGILANTHUS (LORANTHACEAE)
cuneifolius - *Citheronia vogleri.*
eugenioides - *Citheronia vogleri.*

PHYLLANTHUS (EUPHORBIACEAE)
sp. unspecified - *Coscinocera hercules.*
emblica - *Attacus atlas.*
velutinus - *Attacus atlas.*

PHYSOCARPUS (ROSACEAE)
opulifolius - *Hyalophora cecropia.*

PICEA (PINACEAE)
sp. unspecified - *Eacles imperialis* ssp., *E. imperialis imperialis, Hyalophora euryalus euryalus.*
glauca - *Eacles imperialis pini, Hyalophora columbia* ssp.
marina - *Hyalophora columbia* ssp.

PIMPINELLA (APIACEAE)
saxifraga - *Eudyaria venata.*

PINUS (PINACEAE)
sp. unspecified - *Antheraea polyphemus* ssp., *Callosamia promethea, Citheronia sepulcralis, Eacles imperialis* ssp., *E. imperialis imperialis, Imbrasia cytherea, I. macrops.*
banksiana - *Eacles imperialis* ssp., *E. imperialis pini.*
caribaea - *Citheronia sepulcralis.*
contorta - *Coloradia pandora pandora, C. pandora lindseyi.*
contorta var. ***latifolia*** - *Coloradia pandora pandora.*
contorta var. ***murrayana*** - *Coloradia pandora lindseyi.*
edulis - ?*Coloradia pandora davisi, C. velda.*

echinata - *Coloradia pandora* ssp., *Eacles imperialis* ssp.
griffithii - *Eacles imperialis* ssp.
halepensis - *Citheronia sepulcralis.*
jeffreyi - *Coloradia pandora lindseyi, C. velda.*
laricio - *Graellsia isabellae* ssp.
maritima - *Graellsia isabellae* ssp.
massoniana - *Actias dubernardi.*
mitis - *Citheronia sepulcralis.*
monophylla - *Coloradia velda.*
mugo - *Citheronia sepulcralis.*
nigra - *Citheronia sepulcralis, Copaxa cydippe.*
patula - *Holocerina smilax, Pseudobunaea irius.*
ponderosa - *Citheronia sepulcralis, Coloradia doris, C. pandora* ssp., *C. pandora davisi, C. pandora lindseyi, Coloradia pandora pandora.*
radiata - *Copaxa cydippe, C. medea, Gonimbrasia belina, Imbrasia cytherea clarki.*
resinosa - *Eacles imperialis pini.*
rigida - *Citheronia sepulcralis, Eacles imperialis imperialis.*
serotina - *Eacles imperialis* ssp.
strobus - *Citheronia sepulcralis, Coloradia doris, C. luski, C. pandora pandora, C. velda, Eacles imperialis pini, E. imperialis* ssp.
sylvestris - *Citheronia sepulcralis, Coloradia pandora davisi, C. pandora lindseyi, C. pandora pandora, C. velda, E. imperialis pini, Graellsia isabellae* ssp., *Imbrasia cytherea.*
taeda - *Eacles imperialis* ssp.
thunbergiana - *Coloradia pandora lindseyi, C. velda.*

PIPER (PIPERACEAE)
sp. unspecified - *Attacus atlas, Ormiscodes cinnamomea.*

PIPTADENIA (FABACEAE: MIMOSOIDEAE)
sp. unspecified - *Aurivillius arata, Automeris amanda* ssp.
buchanani - *Aurivillius arata.*

PISTACIA (ANACARDIACEAE)
chinensis - *Neoris haraldi.*
lentiscus - *Citheronia splendens sinaloensis, Eacles imperialis cacicus.*
terebinthus - *Saturnia pavonia* ssp.

PITHECELLOBIUM (FABACEAE: MIMOSOIDEAE)
lanceolatum - *Attacus atlas, ?Hylesia lineata.*
saman - *Syssphinx colla* ssp., *S. molina, S. quadrilineata.*

PLATANUS (PLATANACEAE)

sp. unspecified - *Aglia tau* ssp., *Antheraea polyphemus* ssp., *Automeris coresus, A. io io, A. liberia, Caligula japonica* ssp., *Dryocampa rubicunda rubicunda, Eacles imperialis* ssp., *E. imperialis imperialis, Gonimbrasia belina, Leucanella memusae* ssp., *L. viridescens* ssp., *Molippa rosea, Rhodinia fugax, Samia cynthia* ssp., *Saturnia pyri.*

lindeniana - *Actias luna* ssp., *Leucanella saturata, Paradirphia hoegei.*

occidentalis - *Antheraea polyphemus* ssp., *Automeris io* ssp., *A. metzli, Eacles imperialis* ssp.

orientalis - *Automeris coresus, A. illustris, A. melanops, Citheronia regalis, Eacles imperialis* ssp., *Hyperchiria incisa* ssp.

PLUMERIA (APOCYNACEAE)

acuminata - *Samia cynthia* ssp.

PODOCARPUS (PODOCARPACEAE)

elata - *Opodiphthera* sp.

spinulosus - *Opodiphthera* sp., *Opodiphthera eucalypti.*

POGA (RHIZOPHORACEAE)

sp. unspecified - *Pseudantheraea arnobia.*

oleosa - *Pseudantheraea discrepans.*

POLYSCIAS (ARALIACEAE)

Polyscias - *Coscinocera hercules.*

fulva - *Nudaurelia anthina.*

POPULUS (SALICACEAE)

sp. unspecified - *Actias luna* ssp., *Archaeoattacus edwardsii, Attacus atlas, Automeris io io, Callosamia promethea, Hemileuca eglanterina* ssp., *Hylesia nigricans, Ormiscodes cinnamomea, Saturnia pyri, S. spini.*

alba - *Perisomena caecigena.*

balsamifera - *Automeris io* ssp., *Hyalophora cecropia.*

deltoides - *Automeris io* ssp., *Caligula japonica* ssp., *C. jonasi, Pseudobunaea irius, Saturnia lindia.*

fremontii - *Hemileuca artemis, H. maia, H. nevadensis.*

gileadensis - *Automeris io* ssp.

grandidentata - *Antheraea polyphemus* ssp.

nigra - *Periphoba hircia, Perisomena caecigena.*

tremuloides - *Actias luna* ssp., *Antheraea polyphemus* ssp., *Automeris io* ssp., *Coloradia pandora* ssp, *Hemileuca maia, H. nevadensis, Hyalophora cecropia.*

PORLIERIA (ZYGOPHYLLACEAE)
lorentzii - *Hylesia remex.*

POTENTILLA (ROSACEAE)
sp. unspecified - *Hemileuca eglanterina* ssp., *Saturnia pavonia* ssp.

POTERIUM (ROSACEAE)
sp. unspecified - *Eudyaria venata.*

PROSOPIS (FABACEAE: MIMOSOIDEAE)
sp. unspecified - *Hemileuca tricolor, Syssphinx hubbardi.*
alba - *Automeris oberthurii.*
algarobilla - *Leucanella aspera* ssp.
glandulosa - *Hemileuca juno, Syssphinx heiligbrodti.*
juliflora - *Hemileuca juno, Syssphinx hubbardi, S. quadrilineata* ssp.
nigra - *Automeris oberthurii.*

PROTEA (PROTEACEAE)
sp. unspecified - *Cinabra hyperbius, Holocerina agomensis, Imbrasia carnegiei.*
affinis - *Imbrasia emini.*
mellifera - *Imbrasia cytherea.*
mundii - *Imbrasia cytherea.*

PRUNUS (ROSACEAE)
sp. unspecified - *Actias artemis, A. luna* ssp., *A. selene* ssp., *Agapema galbina* ssp., *Antheraea mylitta, A. pernyi, A. polyphemus* ssp., *A. yamamai* ssp., *Archaeoattacus edwardsii, Attacus atlas, Automeris beckeri, A. excreta, A. inornata, A. io* ssp., *A. jucunda, A. naranja* ssp., *Bunaea alcinoe, Caligula boisduvali* ssp., *C. japonica* ssp., *Callosamia promethea, Cinabra hyperbius, Cirina forda, Citheronia brissotii* ssp., *Copaxa mazaorum, C. rufinans, Cricula andrei, C. trifenestrata, Dirphia avia, Dirphiopsis delta, Eacles imperialis* ssp., *Gynanisa maja* ssp., *Hemileuca burnsi, H. eglanterina* ssp., *H. maia, H. neumoegeni, H. nevadensis, Hyalophora cecropia H. euryalus euryalus, Imbrasia cleoris, I. gueinzii, I. macrothyris, Leucanella leucane, Paradirphia boudinoti, P. hoegei, P. valverdei, Periphoba hircia, Rhodinia fugax, Rothschildia aurota speculifera, R. lebeau* ssp., *R. orizaba* ssp., *Salassa iris, Samia cynthia* ssp., *Saturnia atlantica, S. pavonia* ssp., *S. pyri, S. spini.*
americana - *Hyalophora cecropia.*
amygdalus - *Actias selene* ssp., *Attacus atlas, Neoris* sp., *Saturnia pyri.*
armeniaca - *Actias artemis, A. selene ningpoana, Adetomeris erythrops, Attacus atlas, Caligula japonica* ssp., *Rothschildia hesperus* ssp., *R. orizaba* ssp., *Saturnia atlantica, S. pyri.*
avium - *Hemileuca eglanterina* ssp., *Hyalophora cecropia, Saturnia lindia.*

capuli - *Automeris* sp., *Leucanella contempta contempta.*
cerasus - *Actias selene* ssp., *Antheraea polyphemus* ssp., *Automeris io* ssp., *Hemileuca eglanterina* ssp., *H. maia, Hyalophora cecropia, H. columbia* ssp.
demissa - *Hyalophora columbia gloveri, H. euryalus euryalus.*
domestica - *Actias luna* ssp., *Antheraea paphia, A. polyphemus* ssp., *Automeris io* ssp., *Callosamia promethea, Citheronia regalis, Hemileuca eglanterina* ssp., *Hyalophora columbia columbia, H. columbia gloveri, Saturnia pavonia* ssp.
domestica var. **galatensis** - *Antheraea polyphemus* ssp., *Hemileuca eglanterina* ssp., *Hyalophora cecropia.*
emarginata - *Hemileuca eglanterina* ssp., *Hyalophora columbia gloveri, H. euryalus euryalus.*
fasciculata - *Hemileuca burnsi, H. neumoegeni.*
ilicifolia - *Hyalophora cecropia.*
maritima - *Hyalophora cecropia.*
padus - *Actias selene* ssp., *Saturnia lindia.*
pensylvanica - *Automeris io* ssp., *Callosamia promethea, Hyalophora cecropia, H. columbia* ssp.
persica - *Antheraea polyphemus* ssp., *Attacus atlas, Caligula japonica* ssp., *Callosamia promethea, Citheronia brissotii meridionalis, Dirphia curitiba, Gynanisa maja* ssp., *Imbrasia wahlbergi, Neoris haraldi, Rothschildia lebeau forbesi, Saturnia pyri.*
pseudo-cerasus - *Actias artemis, A. selene ningpoana.*
puddum - *Antheraea knyvetti, A. roylei.*
salicina - *Caligula japonica* ssp.
serotina - *Antheraea polyphemus* ssp., *Automeris io* ssp., *A. io io, A. io lilith, A. louisiana, A. randa, A. zephyria, Callosamia angulifera, C. promethea, Citheronia regalis, Coscinocera hercules, Eacles imperialis opaca, E. imperialis* ssp., *Eupackardia calleta, Hemileuca eglanterina* ssp., *H. lucina, H. maia, Hyalophora cecropia, H. columbia* ssp., *H. columbia gloveri, H. euryalus euryalus, Rothschildia cincta* ssp., *R. jorulla* ssp., *R. lebeau forbesi, R. jacobaeae, R. lebeau aroma, R. orizaba* ssp., *R. roxana, Samia cynthia* ssp., *Saturnia pavonia* ssp.
spinosa - *Attacus atlas, Saturnia pavonia* ssp., *S. pyri, S. spini.*
triflora - *Automeris complicata.*
virginiana - *Actias artemis, Antheraea polyphemus* ssp., *Automeris io* ssp., *A. io lilith, Caligula japonica* ssp., *Callosamia promethea, Eacles imperialis* ssp., *E. imperialis imperialis, Hemileuca burnsi, H. eglanterina* ssp., *H. neumoegeni, Hyalophora cecropia, H. columbia columbia, H. columbia* ssp., *H. columbia gloveri, H. euryalus euryalus, Rothschildia jacobaeae, R. orizaba* ssp., *Saturnia pavonia* ssp.
vulgaris - *Hyalophora cecropia.*

PSEUDOTSUGA (PINACEAE)
menziesii - *Hyalophora columbia columbia, H. columbia gloveri, H. euryalus euryalus.*

PSIDIUM (MYRTACEAE)

sp. unspecified - *Holocerina smilax, Imbrasia cytherea* ssp., *I. dione, I. wahlbergi, Othorene purpurascens.*

albortum - *Citheronia laocoon.*

araca - *Citheronia laocoon.*

cujavillus - *Attacus lorquinii.*

guajava - *Attacus lorquinii, Automeris complicata, Citheronia laocoon, C. lobesis* ssp., *Cricula trifenestrata, Eriogyna pyretorum, Gamelia abasia, Holocerina smilax, Imbrasia wahlbergi.*

guineense - *Hylesia lineata.*

incanescens - *Citheronia laocoon.*

mediterraneum - *Citheronia laocoon.*

petrosum - *Citheronia laocoon.*

pomiferum - *Citheronia laocoon, Eacles penelope.*

pyriforme - *Attacus lorquinii.*

PSOROSPERMUM (HYPERICACEAE)

sp. unspecified - *Holocerina angulata.*

PTEROCARPUS (FABACEAE: FABOIDEAE)

sp. unspecified - *Gynanisa maja* ssp., *Pseudobunaea epithyrena.*

delevoyi - *Pseudobunaea irius.*

PTEROCARYA (JUGLANDACEAE)

stenoptera - *Actias selene ningpoana, Eriogyna pyretorum.*

PUNICA (PUNICACEAE)

sp. unspecified - *Actias artemis.*

PURSHIA (ROSACEAE)

sp. unspecified - *Hyalophora columbia gloveri.*

tridentata - *Hemileuca eglanterina* ssp., *H. nuttalli nuttalli, H. nuttalli uniformis, Hyalophora columbia gloveri, H. euryalus euryalus.*

PYCREUS (CYPERACEAE)

muricatus - *Bunaeopsis oubie.*

PYRACANTHA (ROSACEAE)

sp. unspecified - *Hemileuca eglanterina* ssp., *H. nuttalli nuttalli.*

PYRUS (ROSACEAE)

sp. unspecified - *Attacus atlas, Caligula boisduvali* ssp., *Neoris* sp., *Saturnia atlantica.*

americana - *Hyalophora cecropia.*

calleryana - *Saturnia lindia*.

communis - *Actias artemis*, *A. selene* ssp., *A. selene ningpoana*, *Aglia tau* ssp., *Antheraea polyphemus* ssp., *A. roylei*, *Attacus atlas*, *Automeris complicata*, *A. io* ssp., *Caligula boisduvali* ssp., *C. japonica* ssp., *C. simla*, *Cricula andrei*, *C. trifenestrata*, *Hyalophora cecropia*, ?*Imbrasia cytherea*, *Neoris haraldi*, *Perisomena caecigena*, *Saturnia pavonia* ssp., *S. pyri*.

coronaria - *Hyalophora cecropia*.

malus - *Actias artemis*, *A. selene* ssp., *Aglia tau* ssp., *Antheraea assamensis*, *A. pernyi*, *A. roylei*, *A. yamamai* ssp., *Attacus atlas*, *Automeris banus* ssp., *A. excreta*, *A. io* ssp., *A. io coloradensis*, *A. janus*, *A. liberia*, *A. metzli*, *A. zephyria*, *Bunaea alcinoe*, *Caligula japonica* ssp., *C. simla*, *Callosamia promethea*, *Cirina forda*, *Cricula andrei*, *C. trifenestrata*, *Eupackardia calleta*, *Gonimbrasia tyrrhea* ssp., *Hemileuca eglanterina* ssp., *Holocerina smilax*, *Hyalophora cecropia*, *H. euryalus euryalus*, *Hylesia nigricans*, *Imbrasia cytherea*, *I. gueinzii*, *I. macrothyris*, *Leucanella leucane*, *Neoris haraldi*, *Paradirphia hoegei*, *P. lasiocampina*, *Periphoba hircia*, *Samia cynthia* ssp., *Saturnia atlantica*, *S. pavonia* ssp., *S. pyri*, *S. spini*.

pyrifolia - *Eriogyna pyretorum*.

QUAMOCLIT (CONVOLVULACEAE)

setosa - *Hyperchiria orodina*.

QUERCUS (FAGACEAE)

sp. unspecified - *Actias artemis*, *A. luna* ssp., *A. maenas*, *A. selene* ssp., *Anisota dissimilis*, *A. stigma* ssp., *A. virginiensis*, *Antheraea assamensis*, *A. godmani*, *A. knyvetti*, *A. montezuma*, *A. mylitta*, *A. paphia*, *A. polyphemus* ssp., *A. roylei*, *A. yamamai* ssp., *Attacus lorquinii*, *Automerella flexuosa*, *Automeris beckeri*, *boudinotiana*, *A. excreta*, *A. io* ssp., *A. io io*, *A. inornata*, *A. iris hesselorum*, *A. naranja* ssp., *A. randa*, *A. tridens*, *Bunaeopsis licharbas*, *Caligula boisduvali* ssp., *C. boisduvali fallax*, *C. japonica* ssp., ?*C. simla*, *Callosamia promethea*, *Citheronia regalis*, *Copaxa lavendera*, *C. muellerana*, *C. rufinans*, *Cricula andrei*, *C. trifenestrata*, *Dirphia curitiba*, *Eacles ducalis*, *E. imperialis decoris*, *E. imperialis* ssp., *E. imperialis imperialis*, *E. ormondei* ssp., *Hemileuca eglanterina* ssp., ?*H. lucina*, *H. rubridorsa*, *H. tricolor*, *Holocerina smilax*, *Hyalophora euryalus euryalus*, *Hylesia continua cressida*, *H. lineata*, *H. nigricans*, *Hyperchiria incisa* ssp., *Imbrasia cytherea*, *I. wahlbergi*, *Lonomia cynira*, *Melanocera menippe*, *Paradirphia lasiocampina*, *Perisomena caecigena*, *Pseudimbrasia deyrollei*, *Rhodinia fugax*, *Rothschildia arethusa* ssp., *R. cincta* ssp., *R. jorulla* ssp., *Saturnia pavonia* ssp., *Syssphinx bicolor*.

X bebbiana - *Antheraea polyphemus* ssp.

acuta - *Caligula cachara*.

acutissima - *Antheraea yamamai superba*, *Antheraea yamamai* ssp., *Caligula japonica arisana*, *Rhodinia fugax*, *R. verecunda*.

agrifolia - *Adeloneivaia jason* ssp., *Copaxa mannana, Hemileuca nevadensis.*
alba - *Actias luna* ssp., *Anisota finlaysoni, A. senatoria, A. stigma stigma, A. virginiensis, Antheraea mylitta, A. polyphemus* ssp., *Automeris metzli, A. zephyria, Hyalophora cecropia.*
albocincta - *Anisota assimilis.*
aliena - *Rhodinia verecunda.*
arizonica - *Anisota oslari, Hemileuca lex.*
bicolor - *Anisota senatoria.*
cerris - *Perisomena caecigena.*
chihuahuaensis - *Eacles imperialis oslari.*
coccinea - *Anisota senatoria, Antheraea polyphemus* ssp., *Automeris metzli, Dryocampa rubicunda* ssp., *Eacles imperialis* ssp.
dealbata - *Antheraea roylei.*
dentata - *Aglia tau* ssp., *Antheraea yamamai* ssp., *Rhodinia fugax.*
dilatata - *Antheraea roylei.*
dumosa - *Adeloneivaia jason* ssp.
emoryi - *Automeris pamina, Eacles imperialis oslari, Hemileuca diana.*
falcata - *Anisota consularis, A. peigleri, A. pellucida discolor, A. pellucida pellucida, A. stigma fuscosa, Antheraea polyphemus* ssp.
fusiformis - *Hemileuca slosseri.*
gambelii - *Anisota peigleri, A. oslari, Automeris io coloradensis, A. pamina, A. randa, A. zephyria, Copaxa mannana, Eacles imperialis oslari, Gynanisa maja* ssp., *Hemileuca diana, H. grotei, Holocerina smilax, Othorene verana.*
grisea - *Anisota assimilis, Automeris cecrops peigleri.*
havardii - *Hemileuca peigleri, H. slosseri.*
havardii x ***Q. stellata*** - *Hemileuca grotei.*
himalayana - *Antheraea roylei.*
ilex - *Actias selene ningpoana, Anisota pellucida pellucida, Antheraea roylei, Caligula cachara, Copaxa mannana, Dirphia avia, D. baroma, Eacles imperialis cacicus, E. penelope, Hyperchiria incisa* ssp., *Periphoba hircia, Perisomena caecigena, ?Pseudodirphia eumedide.*
ilicifolia - *Anisota senatoria, Dryocampa rubicunda* ssp., *Hemileuca maia.*
imbricaria - *Anisota pellucida pellucida, A. senatoria.*
incana - *Antheraea roylei.*
kelloggii - *Adeloneivaia jason* ssp., *Anisota oslari, Copaxa mannana.*
laevis - *Anisota consularis, Dryocampa rubicunda rubicunda, Eacles imperialis* ssp., *Hemileuca maia.*
lobata - *Hemileuca diana.*
lyrata - *Anisota pellucida pellucida, Antheraea polyphemus* ssp.
macrocarpa - *Anisota manitobensis, A. pellucida discolor, A. senatoria, A. virginiensis, Antheraea polyphemus* ssp., *Eacles imperialis* ssp., *Hemileuca slosseri.*

marilandica - *Anisota peigleri, A. pellucida discolor, A. pellucida pellucida, A. stigma fuscosa, Automeris pamina, Hemileuca grotei.*

michauxii - *Anisota stigma stigma.*

muehlenbergii - *Anisota finlaysoni.*

myrtifolia - *Anisota consularis, A. pellucida pellucida.*

nigra - *Anisota consularis, A. peigleri, A. pellucida discolor, A. pellucida pellucida, A. stigma fuscosa, A. stigma stigma, A. virginiensis, Automeris zephyria, Antheraea polyphemus* ssp., *Perisomena caecigena.*

nuttalli - *Anisota pellucida pellucida.*

oblongifolia - *Anisota oslari, Automeris zephyria, Eacles imperialis oslari, Hemileuca diana.*

oleoides - *Automeris rubrescens* ssp., *A. zugana, Othorene verana, Periphoba arcaei.*

palustris - *Anisota consularis, A. finlaysoni, A. oslari, A. peigleri, A. pellucida discolor, A. pellucida pellucida, A. senatoria, A. stigma fuscosa, A. stigma stigma, A. virginiensis, Antheraea polyphemus* ssp., *Automeris io* ssp., *A. iris hesselorum, A. pamina, A. randa, A. zephyria, Eacles imperialis pini, E. imperialis* ssp., *Hemileuca diana.*

pedunculata - *Holocerina smilax.*

petraea - *Anisota virginiensis, Automeris zephyria, Perisomena caecigena.*

phellos - *Antheraea pernyi, Automeris zephyria.*

prinoides - *Anisota senatoria, Hemileuca maia.*

prinus - *Anisota finlaysoni, A. senatoria, A. virginiensis.*

pubescens - *Anisota peigleri, A. pellucida discolor, A. pellucida pellucida, A. stigma stigma, Perisomena caecigena.*

robur - *Aglia tau* ssp., *Anisota virginiensis, Antheraea harti, A. mylitta, A. pernyi, A. polyphemus* ssp., *A. roylei, A. yamamai* ssp., *Automeris coresus, A. rubrescens* ssp., *A. zephyria, A. zozine, Bunaea alcinoe, Copaxa mannana, Eacles imperialis anchicayensis, E. imperialis opaca, E. imperialis oslari, Gonimbrasia tyrrhea* ssp., *Gynanisa maja, Hemileuca diana, H. peigleri, H. slosseri, Perisomena caecigena, Saturnia lindia.*

rubra - *Actias luna* ssp., *Anisota finlaysoni, A. oslari, A. peigleri, A. senatoria, A. stigma stigma, A. virginiensis, Antheraea polyphemus* ssp., *Automeris io neomexicana, A. pamina, A. randa, A. zephyria, Eacles imperialis oslari, E. imperialis* ssp., *Hemileuca tricolor.*

rubra var. **ambigua** - *Hemileuca maia.*

semicarpifolia - *Antheraea roylei.*

semiserrata - *Antheraea roylei, A. yamamai* ssp.

serrata - *Aglia tau* ssp., *Antheraea pernyi, A. roylei, Rhodinia fugax.*

shumardii - *Hemileuca maia.*

stellata - *Anisota pellucida discolor, A. stigma fuscosa, A. virginiensis, Antheraea polyphemus* ssp., *Hemileuca grotei, H. maia.*

texana - *Hemileuca grotei, H. peigleri.*

turbinella - *Anisota oslari, Hemileuca diana.*

turneri - *Antheraea roylei, A. semperi, Gynanisa maja.*
undulata - *Automeris pamina, Hemileuca diana.*
variabilis - *Caligula thibeta okurai.*
velutina - *Anisota peigleri, A. oslari, A. pellucida discolor, A. pellucida pellucida, A. senatoria, A. stigma stigma, A. virginiana, Antheraea polyphemus* ssp., *Dryocampa rubicunda* ssp., *Hemileuca maia.*
virens - *Antheraea polyphemus* ssp.
virginiana - *Anisota consularis, A. stigma stigma, Automeris louisiana, Hemileuca maia, H. peigleri.*
virginiana var. ***fusiformis*** - *Hemileuca grotei.*

QUISQUALIS (COMBRETACEAE)
indica - *Attacus atlas.*

RHAMNUS (RHAMNACEAE)
sp. unspecified - *Adetomeris erythrops, Antheraea polyphemus* ssp., *Hyalophora cecropia, Imbrasia cytherea* ssp., *Saturnia atlantica.*
alnifolia - *Hemileuca eglanterina* ssp.
californicus - *Agapema homogena, Hemileuca eglanterina* ssp.
californicus var. ***tomentella*** - *Hyalophora euryalus euryalus.*
crocea - *Hyalophora euryalus euryalus.*
frangula - *Actias selene* ssp., *Agapema homogena, Epiphora mythimnia.*
prunoides - *Imbrasia cytherea.*
purshiana - *Hemileuca eglanterina* ssp., *Hyalophora euryalus euryalus.*

RHAPIS (ARECACEAE)
sp. unspecified - *Automeris io lilith.*

RHIZOPHORA (RHIZOPHORACEAE)
sp. unspecified - *Automeris io lilith.*
calceolaris - *Antheraea paphia.*
mangle - *Automeris io* ssp.

RHODODENDRON (ERICACEAE)
sp. unspecified - *Actias selene* ssp., *Attacus atlas, Automeris io* ssp., *A. io lilith, Callosamia promethea, Cricula trifenestrata.*
canadense - ?*Hyalophora columbia columbia.*
ponticum - *Cricula andrei.*

RHUS (ANACARDIACEAE)
sp. unspecified - *Citheronia regalis, C. splendens* ssp., *Eacles imperialis* ssp., *E. imperialis imperialis, G. zambesina* ssp., *Hyalophora euryalus euryalus.*
angustifolia - *Imbrasia cytherea.*

aromatica var. **flabelliformis** - *Hemileuca chinatiensis.*
choriophylla - *Citheronia splendens sinaloensis.*
cismontana - *Citheronia regalis.*
copallina - *Actias maenas, Argema mittrei, Citheronia regalis.*
crenata - *Imbrasia cytherea.*
dura - *Imbrasia cytherea clarki.*
glabra - *Actias luna* ssp., *A. maenas, Bunaea alcinoe, Citheronia regalis, Eacles imperialis* ssp., *Holocerina smilax, Hyalophora cecropia.*
integrifolia - *Hyalophora euryalus cedrosensis, Saturnia mendocino, S. walterorum.*
laevigata - *Holocerina smilax.*
lancea - *Gonimbrasia belina.*
laurina - *Actias selene* ssp., *Adetomeris erythrops, Argema mimosae, A. mittrei, Automeris banus* ssp., *A. phrynon, A. pamina, A. zozine, Cirina forda, Citheronia azteca* ssp., *C. beledonon, C. bellavista* ssp., *C. regalis, C. splendens sinaloensis, Copaxa lavendera, C. rufinans, Eacles imperialis oslari, E. imperialis* ssp., *E. masoni* ssp., *E. ormondei* ssp., *Eupackardia calleta, Gonimbrasia tyrrhea* ssp., *Hyalophora cecropia, H. euryalus cedrosensis, H. euryalus euryalus, Leucanella acutissima, Lonomia achelous, Rothschildia lebeau* ssp., *R. orizaba* ssp., *Saturnia walterorum.*
longifolia - *Cirina forda, Gonimbrasia belina, Holocerina smilax, Lobobunaea angasana, Nudaurelia belina.*
lucida - *Imbrasia cytherea.*
microphylla - *Hemileuca chinatiensis, H. conwayae.*
radicans - *Actias maenas.*
tomentosa - *Imbrasia cytherea.*
toxicodendron - *Argema mittrei.*
trilobata - *Copaxa lavendera, Hemileuca neumoegeni, Holocerina smilax, Hyalophora columbia gloveri, Saturnia mendocino, S. walterorum.*
typhina - *Actias maenas, Argema mittrei, Citheronia azteca* ssp., *C. beledonon, C. regalis, C. splendens sinaloensis, Hyalophora cecropia, Lemaireia luteopeplus aureopeplus, Samia cynthia* ssp., *Saturnia mendocino.*
verniciflua - *Caligula japonica* ssp.
villosa - *Holocerina smilax.*

RHYNCHOSIA (FABACEAE: FABOIDEAE)
reticulata - *Automeris io io.*

RIBES (SAXIFRAGACEAE)
sp. unspecified - ?*Agapema galbina* ssp., ?*Antheraea polyphemus* ssp., *Automeris io* ssp., *Hyalophora columbia gloveri, Hyalophora euryalus euryalus.*
alpinum - *Agapema homogena, Hyalophora cecropia, H. columbia gloveri, H. euryalus euryalus.*
americanum - *Hyalophora cecropia.*

cereum - *Agapema homogena.*
cynosbati - *Antheraea polyphemus* ssp.
floridum - *Hyalophora cecropia.*
floridum var. ***officinale*** - *Hyalophora cecropia.*
grossularia - *Hyalophora cecropia.*
lacustre - *Hyalophora columbia gloveri.*
nigrum - *Hyalophora cecropia.*
rubrum - *Hyalophora cecropia.*
rubrum var. ***officinale*** - *Hyalophora cecropia.*
sanguineum - *Antheraea polyphemus* ssp., *Hyalophora euryalus euryalus.*
sativum - *Antheraea polyphemus* ssp., *Automeris io* ssp., *Hyalophora cecropia, H. columbia gloveri.*

RICINUS (EUPHORBIACEAE)
communis - *Antheraea paphia, Attacus atlas, Citheronia laocoon, Imbrasia done, I. wahlbergi, Rothschildia aurota* ssp., *Samia cynthia ricini, S. cynthia* ssp., *S. cynthia walkeri.*

ROBINIA (FABACEAE: FABOIDEAE)
sp. unspecified - *Automeris io* ssp., *A. iris iris, A. naranja, Eacles imperialis* ssp.
neomexicana - *Automeris pamina, Hemileuca tricolor.*
pseudoacacia - *Automerella flexuosa, Automeris banus* ssp., *A. excreta, A. hamata, A. incarnata, A. inornata, A. io* ssp., *A. io draudtiana, A. io io, A. jucunda, A. leucane, A. iris hesselorum, A. liberia, A. montezuma, A. moresca, A. phrynon, A. zozine, Cinabra hyperbius, Citioica anthonilis, Dirphia panamensis, Eacles imperialis oslari, Gynanisa maja* ssp., *Hemileuca lex, H. rubridorsa, Hylesia lineata, Hyperchiria nausica, Imbrasia ertli, I. oyemensis, Leucanella aspera* ssp., *L. memusae* ssp., *Lonomia cynira, Ludia delegorguei, Molippa nibasa, M. rosea, Paradirphia boudinoti, P. hoegei, P. lasiocampina, P. semirosea, Periphoba hircia, Pseudodirphia eumedide, P. eumedidoides, Syssphinx albolineata, S. bicolor, S. bisecta, S. gomezi, S. hubbardi, S. molina, S. montana, Urota sinope.*

ROSA (ROSACEAE)
sp. unspecified - *Actias artemis, Antheraea polyphemus* ssp., *Attacus atlas, Automeris io* ssp., *A. io io, A. io lilith, A. muscula, Bunaea alcinoe, Caligula boisduvali* ssp., *Citheronia laocoon, Eacles imperialis* ssp., *Hemileuca eglanterina* ssp., *Hyalophora cecropia, H. columbia gloveri, H. euryalus euryalus, Imbrasia cleoris, Pseudobunaea illustris, Saturnia atlantica, S. pavonia* ssp., *S. spini.*
blanda - *Hyalophora cecropia.*
multifolia - *Eriogyna pyretorum.*

ROTTBOELLIA (POACEAE)
compressa var. ***fasciculata*** - *Pseudautomeris grammivora.*

ROUREA (CONNARACEAE)

glabra - *Automeris io lilith, Periphoba arcaei.*

RUBUS (ROSACEAE)

sp. unspecified - *Hemileuca eglanterina* ssp., *Hyalophora cecropia, Saturnia pavonia* ssp., *S. pyri, S. spini.*

allegheniensis - *Anisota senatoria, Automeris io* ssp., *Hemileuca eglanterina* ssp., *Hyalophora cecropia.*

canadensis - *Automeris io* ssp.

idaeus - *Saturnia pavonia* ssp.

idaeus var. **strigosus** - *Anisota senatoria, Automeris io* ssp., *Hyalophora cecropia.*

occidentalis - *Hyalophora cecropia.*

ulmifolius - *Adetomeris erythrops.*

RUMEX (POLYGONACEAE)

sp. unspecified - *Saturnia pavonia* ssp.

SALIX (SALICACEAE)

sp. unspecified - *Actias artemis, A. luna* ssp., *A. selene* ssp., *A. selene ningpoana, Agapema homogena, Antheraea paphia, A. pernyi, A. polyphemus* ssp., *Antherina suraka, Archaeoattacus edwardsii, Attacus atlas, A. caesar, A. lorquinii, Aurivillius arata, Automeris hamata, A. io* ssp., *A. io io, A. io coloradensis, A. io neomexicana, A. naranja* ssp., *A. rubrescens* ssp., *A. zephyria, Bunaeopsis oubie, Caligula japonica* ssp., *Callosamia promethea, Citheronia hamifera* ssp., *C. regalis, Copaxa decrescens, Cricula andrei, Dirphia avia, D. panamensis, Eacles imperialis* ssp., *E. penelope, Eupackardia calleta, Gonimbrasia tyrrhea* ssp., *Hemileuca eglanterina* ssp., ?*H. lucina, H. maia, H. rubridorsa, Hyalophora euryalus euryalus, H. columbia gloveri, Hylesia nigricans, Imbrasia nictitans, I. rectilineata, Neoris* sp., *Nudaurelia belina, Rhodinia fugax, R. newara, Rothschildia jorulla* ssp., *R. lebeau forbesi, R. lebeau* ssp., *R. orizaba* ssp., *Samia cynthia* ssp., *Saturnia lindia, S. pavonia* ssp., *S. pyri, S. spini.*

alba - *Antheraea polyphemus* ssp., *Hyalophora cecropia, H. columbia* ssp., *H. columbia nokomis.*

babylonica - *Actias selene* ssp., *Antheraea polyphemus* ssp., *Automeris louisiana, A. zephyria, Caligula simla, Copaxa lavendera, Cricula trifenestrata, Eacles imperialis opaca, Eupackardia calleta, Hemileuca maia, H. nevadensis, Hyalophora columbia columbia, H. columbia gloveri, H. euryalus euryalus, Periphoba hircia, Pseudodirphia eumedidoides, Rhodinia newara, Saturnia pavonia* ssp., *S. spini.*

caprea - *Actias luna* ssp., *Adelocephala subangulata* ssp., *Aglia tau* ssp., *Automeris io neomexicana, Bunaeopsis licharbas, Citheronia hamifera* ssp., *Citioica anthonilis, Copaxa multifenestrata* ssp., *Syssphinx molina.*

chilensis - *Paradirphia hoegei.*

discolor - *Hemileuca eglanterina* ssp., *H. nevadensis, Hyalophora cecropia, H. columbia gloveri, H. euryalus euryalus, Rothschildia jorulla* ssp., *R. lebeau forbesi.*
elegans - *Rhodinia newara.*
exigua - *Hyalophora columbia gloveri, H. euryalus euryalus.*
fragilis - *Actias selene* ssp., *Antheraea polyphemus* ssp., *Caligula simla.*
hindsiana - *Hemileuca maia.*
humilis - *Antheraea polyphemus* ssp., *Hyalophora cecropia.*
interior - *Antheraea polyphemus* ssp., *Automeris io* ssp., *Hemileuca maia, H. nevadensis, Hyalophora cecropia, Rothschildia lebeau forbesi.*
lasiandra - *Hemileuca nevadensis.*
lasiolepis - *Hemileuca nevadensis.*
longifolia - *Hemileuca artemis, H. nevadensis.*
lucida - *Hyalophora cecropia.*
viminalis - *Hyalophora cecropia.*

SALVIA (LAMIACEAE)
runcinata - *Ludia delegorguei.*

SAMBUCUS (CAPRIFOLIACEAE)
sp. unspecified - *Antheraea polyphemus* ssp., *Samia cynthia* ssp., *Saturnia pavonia* ssp.
canadensis - *Eacles imperialis* ssp., *Hyalophora cecropia.*
pubens - *Hyalophora cecropia.*
racemosa var. ***pubens*** - *Hyalophora cecropia.*

SANDORICUM (MELIACEAE)
koetjape - *Attacus atlas, A. caesar, A. lorquinii.*

SANGUISORBA (ROSACEAE)
sp. unspecified - *Eudyaria venata.*

SAPINDUS (SAPINDACEAE)
saponaria - *Eacles imperialis pini.*

SAPIUM (EUPHORBIACEAE)
biloculare - *Eupackardia calleta, Rothschildia jorulla* ssp.
discolor - *Samia cynthia walkeri.*
insigne - *Attacus atlas.*
sebiferum - *Actias selene* ssp., *A. selene ningpoana, Antheraea frithi javanensis, Attacus atlas, Samia cynthia walkeri.*

SARCOSTEMMA (ASCLEPIADACEAE)
brevistigma - *Antheraea assamensis, Attacus atlas.*

SASSAFRAS (LAURACEAE)
sp. unspecified - *Automeris io io.*
albidum - *Antheraea polyphemus* ssp., *Automeris io* ssp., *Attacus atlas, Automeris io io, Callosamia angulifera, C. promethea, Eacles imperialis imperialis, Hyalophora cecropia, Rothschildia orizaba* ssp., *Samia cynthia* ssp.
albidum var. **molle** - *Automeris io* ssp., *Citheronia regalis, Eacles imperialis* ssp., *Hyalophora cecropia.*

SAURAUIA (DILLENIACEAE)
sp. unspecified - *Loepa sikkima.*

SCHEFFLERA (ARALIACEAE)
octophylla - *Attacus atlas.*

SCHIMA (THEACEAE)
wallichii - *Actias maenas.*

SCHINUS (ANACARDIACEAE)
sp. unspecified - *Antherina suraka.*
dependens - *Citheronia splendens* ssp., *C. vogleri, Eacles imperialis tucumana.*
gracilipes - *Dirphiopsis ursina.*
molle - *Argema mittrei, Cirina forda, Citheronia splendens splendens, C. splendens sinaloensis, Eacles imperialis oslari, Eupackardia calleta, Hyalophora euryalus cedrosensis, H. euryalus euryalus, Imbrasia krucki, I. wahlbergi, Opodiphthera eucalypti, Rothschildia cincta* ssp., *R. orizaba* ssp., *Usta terpsichore.*
terebinthifolius - *Automeris melanops, Citheronia andina, C. azteca* ssp., *C. beledonon, C. laocoon, Dirphia panamensis, Hylesia acuta, Rothschildia lebeau* ssp., *Saturnia pyri, S. walterorum.*

SCHIZOLOBIUM (FABACEAE: CAESALPINIOIDEAE)
parahybum - *Automeris* sp.

SCHLEICHERA (SAPINDACEAE)
oleosa - *Attacus atlas, Cricula trifenestrata.*
trijugata - *Caligula japonica* ssp.

SCHOTIA (FABACEAE: CAESALPINIOIDEAE)
sp. unspecified - *Gynanisa maja* ssp.
brachypetala - *Gonimbrasia belina.*
speciosa - *Cirina forda.*

SCLERIA (CYPERACEAE)
rehmanii - *Bunaeopsis jacksoni.*

SCLEROCARYA (ANACARDIACEAE)
sp. unspecified - *Usta terpsichore.*
birrea - *Argema mimosae* ssp., *Gonimbrasia belina.*
caffra - *Argema besanti, A. mimosae* ssp., *Gonimbrasia belina, ?Urota sinope, Usta terpsichore.*

SEBASTIANIA (EUPHORBIACEAE)
klotzschiana - *Citheronia laocoon.*

SELINUM (APIACEAE)
sp. unspecified - *Automeris amanda* ssp.

SENECIO (ASTERACEAE)
deltoideus - *Ludia delegorguei.*

SERENOA (ARECACEAE)
repens - *Automeris io* ssp., *A. io lilith.*

SERJANIA (SAPINDACEAE)
sp. unspecified - *Hyperchiria incisa* ssp.
atrolineata - *Automeris zugana.*
schiedeana - *Hylesia lineata.*

SESBANIA (FABACEAE: FABOIDEAE)
grandiflora - *Attacus atlas.*

SETARIA (POACEAE)
viridis - *Attacus atlas.*

SHEPHERDIA (ELAEAGNACEAE)
sp. unspecified - *Hyalophora columbia gloveri.*
argentea - *Hyalophora columbia gloveri, H. euryalus euryalus.*
canadensis - *Hyalophora euryalus euryalus.*

SHOREA (DIPTEROCARPACEAE)
glauca - *?Antheraea larissa.*
robusta - *Antheraea paphia, A. frithi* ssp.
talura - *Antheraea paphia.*

SIDA (MALVACEAE)
vera - *Automeris rubrescens* ssp.

SIDEROXYLON (SAPOTACEAE)
inerme - *Cirina forda*.

SIMMONDSIA (BUXACEAE)
chinensis - *Hemileuca electra* ssp.

SITANION (POACEAE)
jubatum - *Hemileuca oliviae*.

SOLANUM (SOLANACEAE)
hazenii - *Automeris zugana*.
tuberosum - *Leucanella memusae* ssp.
variabile - *Leucanella memusae* ssp.

SORBUS (ROSACEAE)
sp. unspecified - *Antheraea polyphemus* ssp., *Caligula boisduvali* ssp., *Saturnia pavonia* ssp.
americana - *Hyalophora cecropia*.

SORGHUM (POACEAE)
sp. unspecified - *Hemileuca oliviae*.

SOROCEA (MORACEAE)
ilicifolia - *Automeris beckeri*.

SPARTIUM (FABACEAE: FABOIDEAE)
junceum - *Automeris complicata*.

SPATHODEA (BIGNONIACEAE)
campanulata - *Attacus atlas*, *Holocerina angulata*.

SPIRAEA (ROSACEAE)
sp. unspecified - *Automeris io* ssp., *Neoris* sp.
alba - *Saturnia pavonia* ssp.
betulifolia var. **corymbosa** - *Hyalophora cecropia*.
latifolia - *Hemileuca lucina*, *Hyalophora cecropia*.
menziesii - *Hemileuca eglanterina* ssp.
salicifolia - *Hemileuca lucina*, *Hyalophora cecropia*.
tomentosa - *Hemileuca lucina*, *Hyalophora cecropia*, *H. columbia* ssp.

SPIROSTACHYS (EUPHORBIACEAE)
africana - *Argema besanti, A. mimosae* ssp.

SPONDIAS (ANACARDIACEAE)
sp. unspecified - *Cricula trifenestrata.*
lutea - *Imbrasia dione.*
mombin - *Citheronia lobesis* ssp., *Periphoba arcaei, Rothschildia lebeau* ssp.
purpurea - *Rothschildia lebeau* ssp.

SPOROBOLUS (POACEAE)
sp. unspecified - *Bunaeopsis zaddachi.*
airoides - *Hemileuca oliviae.*

STACHYTARPHETA (VERBENACEAE)
cayennensis - *Attacus atlas.*

STIGMAPHYLLON (MALPIGHIACEAE)
ellipticum - *Hylesia lineata.*

STROPHANTHUS (APOCYNACEAE)
sp. unspecified - *Antherina suraka.*

STYRAX (STYRACACEAE)
americana - *Callosamia promethea.*

SWIETENIA (MELIACEAE)
macrophylla - *Automeris* sp.
mahogani - *Attacus atlas, A. caesar.*

SYMPHORICARPOS (CAPRIFOLIACEAE)
sp. unspecified - *Hemileuca eglanterina* ssp., *H. nuttalli uniformis.*
albus - *Automeris io* ssp., *Hemileuca eglanterina* ssp., *H. nuttalli nuttalli, Hyalophora cecropia.*
mollis - *Hemileuca eglanterina* ssp.
oreophilus - *Hemileuca nuttalli nuttalli.*
vaccinoides - *Hemileuca eglanterina* ssp., *H. nuttalli uniformis.*

SYMPLOCOS (SYMPLOCACEAE)
crategoides - *Attacus atlas.*
grandiflora - *Antheraea assamensis.*
tinctoria - *Callosamia promethea.*

SYRINGA (OLEACEAE)

sp. unspecified - *Saturnia pyri*.

vulgaris - *Actias selene* ssp., *Archaeoattacus edwardsii, Attacus atlas, A. caesar, A. lorquinii, Automeris illustris, A. io io, A. naranja* ssp., *A. rubrescens* ssp., *Callosamia promethea, Citheronia regalis, Coscinocera hercules, Eupackardia calleta, Holocerina smilax, Hyalophora cecropia, H. euryalus euryalus, H. columbia gloveri, Leucanella leucane, Ludia delegorguei, Rothschildia arethusa, R. aurota speculifera, R. jacobaeae, R. jorulla* ssp., *R. lebeau forbesi, R. orizaba* ssp., *Samia cynthia* ssp., *Saturnia atlantica*.

SYZYGIUM (MYRTACEAE)

aqueum - *Attacus atlas*.

malaccense - *Attacus atlas*.

owariense - *Imbrasia rectilineata, Micragone cana*.

TABEBUIA (BIGNONIACEAE)

argentea - *Automeris io lilith*.

rosea - *Hylesia lineata*.

TAMARINDUS (FABACEAE: CAESALPINIOIDEAE)

indica - *Automeris melanops*.

TAMARIX (TAMARICACEAE)

gallica - *Hemileuca electra* ssp.

TARCHONANTHUS (ASTERACEAE)

camphoratus - *Ludia delegorguei, L. hansali*.

minor - *Ludia delegorguei*.

TAXODIUM (TAXODIACEAE)

distichum - *Eacles imperialis* ssp.

TECTONA (VERBENACEAE)

grandis - *Antheraea paphia*.

TERMINALIA (COMBRETACEAE)

sp. unspecified - *Actias selene* ssp., *Bunaea alcinoe, Gonimbrasia belina, G. hecate, Nudaurelia belina, Pseudantheraea arnobia, P. discrepans*.

alata - *Antheraea paphia*.

arjuna - *Antheraea paphia*.

bellirica - *Antheraea paphia*.

catappa - *Antheraea paphia, Automeris melanops, Bunaea alcinoe, Citheronia hamifera* ssp., *Citheronia laocoon, Othorene purpurascens*.

crenulata - *Actias selene* ssp.
schweinfurthi - *Gonimbrasia belina, G. osiris annulata, Imbrasia nictitans.*
tomentosa - *Attacus paphia.*
velutina - *Gonimbrasia belina.*

TETRADYMIA (ASTERACEAE)
axillaris - *Hemileuca burnsi.*
glabrata - *Hemileuca burnsi.*
spinosa - *Hemileuca burnsi.*

TETRANTHERA (LAURACEAE)
diglottica - *Antheraea assamensis.*
glauca - *Antheraea assamensis.*
macrophylla - *Antheraea assamensis.*
monopetala - *Antheraea assamensis.*
polyantha - *Antheraea assamensis.*

TETRASTIGMA (VITACEAE)
formosanum - *Loepa katinka formosensis.*

TEUCRIUM (LAMIACEAE)
macrostachyum - *Attacus atlas.*

THEA (THEACEAE)
sinensis - *Attacus atlas.*

THEOBROMA (STERCULIACEAE)
cacao - *Attacus atlas, A. lorquinii, Automerina cypria, A. vala, Imbrasia dione.*

THUJA (CUPRESSACEAE)
occidentalis - *Callosamia promethea.*

TIBOUCHINA (MELASTOMATACEAE)
holosericea - *Hylesia wagneri.*

TILIA (TILIACEAE)
sp. unspecified - *Actias luna* ssp., *Aglia tau* ssp., *Antheraea polyphemus* ssp., *Arsenura rebeli, Automeris io* ssp., *A. io io, A. zephyria, Caio championi, Caligula boisduvali fallax, Callosamia promethea, Eacles imperialis* ssp., *E. imperialis imperialis, Samia cynthia* ssp., *Saturnia pyri.*
americana - *Antheraea polyphemus* ssp., *Automeris io coloradensis, Callosamia promethea, Eacles imperialis* ssp., *Hyalophora cecropia.*
europaea - *Antheraea polyphemus* ssp., *Eacles imperialis* ssp., *Hyalophora cecropia.*

heterophylla - *Eacles imperialis* ssp.
platyphyllos - *Titaea tamerlan*.

TIMONIUS (RUBIACEAE)
sp. unspecified - *Coscinocera hercules*.

TIPUANA (FABACEAE: FABOIDEAE)
speciosa - *Automeris complicata, A. melanops*.

TREMA (ULMACEAE)
bracteolata - *Gonimbrasia belina, Imbrasia wahlbergi, Pseudobunaea tyrrhena*.
guineensis - *Nudaurelia alopia*.
orientalis - *Gonimbrasia belina*.

TRICHILIA (MELIACEAE)
emetica - *Pselaphelia flavivitta*.
lineata - *Pselaphelia flavivitta*.

TRIFOLIUM (FABACEAE: FABOIDEAE)
sp. unspecified - *Automeris io* ssp., *A. io io*.
repens - *Automeris io* ssp.

TRIUMFETTA (TILIACEAE)
lappula - ?*Hylesia lineata*.

TSUGA (PINACEAE)
sp. unspecified - *Eacles imperialis imperialis*.
canadensis - *Eacles imperialis* ssp., *Hyalophora columbia* ssp.

TURNERA (TURNERACEAE)
ulmifolia - *Automeris io lilith*.

TURPINA (STAPHYLEACEAE)
sphaerocarpa - *Actias maenas*.

TURRAEA (MELIACEAE)
floribunda - *Pseudaphelia apollinaris*.
heterophylla - *Pseudaphelia apollinaris*.
nilotica - *Pseudaphelia apollinaris*.

UAPACA (EUPHORBIACEAE)
sp. unspecified - *Argema mittrei, Holocerina agomensis, Pseudantheraea arnobia, P. epithyrena*.

guineensis - *Pseudantheraea discrepans.*
kirkiana - *Holocerina agomensis, Imbrasia rectilineata, I. rubra, Lobobunaea angasana, Ludia delegorguei.*
nitida - *Holocerina agomensis, Imbrasia rubra, Lobobunaea angasana, Pseudobunaea irius, P. pallens.*

ULMUS (ULMACEAE)
sp. unspecified - *Actias luna* ssp., *Antheraea polyphemus* ssp., *Automeris io* ssp., *A. io io, A. zugana, Eacles imperialis* ssp., *E. imperialis imperialis, Gonimbrasia tyrrhea* ssp., *Hyalophora columbia* ssp., *H. columbia gloveri, Neoris* sp., *Saturnia pavonia* ssp.
alata - *Eacles imperialis* ssp.
americana - *Antheraea polyphemus* ssp., *Eacles imperialis* ssp., *Hyalophora cecropia.*
campestris - *Pseudodirphia eumedide, Saturnia pyri, S. spini.*
fulva - *Hyalophora cecropia.*
procera - *Eacles imperialis* ssp.
pumila - *Eriogyna pyretorum.*
racemosa - *Hyalophora cecropia.*
rubra - *Eacles imperialis* ssp., *Hyalophora cecropia.*
thomasii - *Eacles imperialis* ssp., *Hyalophora cecropia.*

UNCARIA (RUBIACEAE)
gambir - *Attacus atlas.*

URENA (MALVACEAE)
sinuata - *Arsenura xanthopus.*

URVILLEA (SAPINDACEAE)
ulmacea - *Hylesia lineata.*

VACCINIUM (ERICACEAE)
sp. unspecified - *Antheraea polyphemus* ssp., *Hyalophora cecropia.*
myrtillus - *Saturnia pavonia* ssp.

VANGUERIA (RUBIACEAE)
spinosa - *Attacus atlas.*

VERNONIA (ASTERACEAE)
natalensis - *Ludia delegorguei.*
oligocephala - *Ludia delegorguei, L. orinoptena.*
polyanthus - *Citheronia laocoon.*
tweedieana - *Citheronia laocoon.*

VIBURNUM (CAPRIFOLIACEAE)
sp. unspecified - *Callosamia promethea.*
cassinoides - *Hyalophora cecropia.*
dentatum - *Automeris io* ssp., *Eacles imperialis* ssp.
opulus var. ***sterile*** - *Caligula japonica* ssp.

VIROLA (MYRISTICACEAE)
bicuhyba - *Rhescyntis hippodamia* ssp.
sebifera - *Automeris* sp.
ucuuba - *Rhescyntis hippodamia* ssp.

VISCUM (LORANTHACEAE)
sp. unspecified - *Opodiphthera loranthi.*

VITIS (VITACEAE)
sp. unspecified - *Antheraea polyphemus* ssp.
californica - *Hemileuca eglanterina* ssp.
inconstans - *Loepa katinka* ssp.
vinifera - *Antherina suraka.*
vulpina - *Hemileuca eglanterina* ssp.

WARBURGIA (CANELLACEAE)
sp. unspecified - *Cirina forda.*

WATSONIA (IRIDACEAE)
iridifolia - *Imbrasia cytherea.*

WEBERA (RUBIACEAE)
corymbosa - *Antheraea paphia.*

WEIGELA (CAPRIFOLIACEAE)
sp. unspecified - *Antheraea polyphemus* ssp.

WEINMANNIA (CUNONIACEAE)
eriorcampa - *Argema mittrei.*

WISTERIA (FABACEAE: FABOIDEAE)
sp. unspecified - *Dirphiopsis ursina, Hyperchiria incisa* ssp.
frutescens - *Automeris io* ssp.
sinensis - *Automeris complicata, A. io* ssp., *Hyalophora cecropia.*

XYLOPIA (ANNONACEAE)
aethiopica - *Micragone nenia, M. nubifera.*

XYLOSMA (FLACOURTIACEAE)
japonicum - *Antheraea pernyi.*

ZANTHOXYLUM (RUTACEAE)
sp. unspecified - *Hylesia lineata.*
acanthopodium - *Actias selene* ssp., *Samia cynthia* ssp.
alatum - *Actias selene* ssp., *Samia cynthia* ssp.
fagara - *Rothschildia lebeau forbesi, R. lebeau* ssp.
hostile - *Samia cynthia* ssp.
setulosum - *Rothschildia lebeau* ssp.

ZEA (POACEAE)
mays - *Automeris io* ssp., *A. io io, A. io lilith, Hemileuca mania.*

ZELKOVA (ULMACEAE)
serrata - *Caligula boisduvali* ssp., *C. japonica* ssp., *Rhodinia fugax.*

ZINNIA (ASTERACEAE)
elegans - *Citheronia laocoon.*

ZIZIPHUS (RHAMNACEAE)
sp. unspecified - *Cricula trifenestrata, Epiphora atbarina.*
jujuba - *Antheraea paphia, Epiphora bauhiniae, Samia cynthia* ssp.
mauretanicus - *Epiphora rectifascia.*
mauritiana - *Epiphora bauhiniae, E. rectifascia.*
mucronata - *Epiphora bauhiniae, E. mythimnia.*
spinachristii - *Epiphora bauhiniae.*

ZUELANIA (FLACOURTIACEAE)
guidonia - *Automeris rubrescens* ssp., *Hylesia lineata, Rothschildia lebeau* ssp.

PUBLISHED FOODPLANT FAMILIES FOR WHICH GENUS AND SPECIES OF PLANT WERE NOT PROVIDED

AMPELIDACEAE - *Goodia nubilata.*

ANACARDIACEAE - *Citheronia laocoon, Nudaurelia belina, Rothschildia maurus.*

APIACEAE - *Eudyaria venata.*

AQUIFOLIACEAE - *Citheronia laocoon.*

AURANTEACEAE - *Rothschildia erycina.*

BIGNONIACEAE - *Rothschildia maurus.*

BURSERACEAE - *Pseudobunaea tyrrhena, Usta terpsichore.*

CARYOCARACEAE - *Citheronia laocoon.*

COMBRETACEAE - *Citheronia laocoon.*

CORNACEAE - *Eacles canaima.*

CUPULIFERAE (Archaic group name: oak [FAGACEAE], hornbeam [CORYLACEAE], etc.) - *Attacus atlas.*

EUPHORBIACEAE - *Citheronia brissotii* ssp., *Imbrasia obscura, Lobobunaea phaedusa, L. saturnus, Nudaurelia anthina, Pseudimbrasia deyrollei, Pseudobunaea cleopatra, Rothschildia maurus, Tagoropsis subocellata.*

FABACEAE - ?*Athletes gigas, Automeris amanda, Micragone ansorgei.*

FABACEAE: CAESALPINIOIDEAE - *Pseudobunaea alinda.*

FABACEAE: MIMOSOIDEAE - *Argema mittrei.*

JUGLANDACEAE - *Actias luna* ssp., *A. truncatipennis, Dirphia curitiba.*

LAMIACEAE - *Ludia orinoptena.*

LAURACEAE - *Archaeoattacus edwardsii, Callosamia promethea.*

LECYTHIDACEAE - *Pseudobunaea tyrrhena.*

MALVACEAE - ?*Arsenura* sp., *Citheronia laocoon, Dysdaemonia boreas, D. fosteri, Eacles ducalis.*

MELASTOMATACEAE - *Arsenura aspasia, Eacles ducalis, E. penelope, Hylesia wagneri.*

MORACEAE - *Pseudimbrasia deyrollei.*

MYRTACEAE - *Citheronia laocoon.*

OCHACEAE - *Melanocera sufferti.*

POACEAE - *Bunaeopsis hersilia.*

RHAMNACEAE - *Imbrasia obscura.*

RHIZOPHORACEAE - *Lobobunaea saturnus.*

ROSACEAE - *Citheronia laocoon, Ormiscodes cinnamomea, Hyalophora cecropia.*

RUBIACEAE - *Homoeopteryx* sp., *Pseudimbrasia deyrollei, Pseudobunaea tyrrhena.*

RUTACEAE - *Rothschildia erycina.*

SAPINDACEAE - *Hyperchiria orodina, Tagoropsis hanningtoni.*

STERCULIACEAE - *Goodia nubilata.*

TILIACEAE - ?*Arsenura* sp., *Dysdaemonia boreas, D. fosteri.*

URTICACEAE - *Automeris beckeri.*

ZINGIBERACEAE - *Goodia nubilata, Ludia obscura.*

VERNACULAR FOODPLANT NAMES FOR WHICH SCIENTIFIC NAMES CANNOT BE FOUND

Bamboo (several genera possible) - *Bunaeopsis hersilia.*
Bean (several general possible) - *Automeris io io, A. io lilith.*
False camphor-tree - *Nudaurelia anthina.*
Fruta de erbe - *Citheronia phoronea.*
Fruit Trees - *Antherina suraka, Cricula andrei, Gynanisa maja, Hyalophora cecropia, Hylesia nigricans, Imbrasia richelmanni, I. rubra, Lobobunaea natalensis, Saturnia pyri.*
Grasses - *Automeris colenon, Bunaeopsis aurantiaca, B. hersilia, B. jacksoni, B. oubie, ?B. phidias, B. zaddachi, Decachorda rosea, Heliconisa pagenstecheri, Hemileuca oliviae.*
Grasses; broad leafed lawn - *Decachorda fulvia.*
Grasses; pasture - *Hemileuca mania.*
Grass; reed-like - *Bunaeopsis arabella, B. jacksoni.*
Locos tree - *Dirphia tarquinia.*
Mimosa; high forest - *Syssphinx molina.*
Mimosa; spiny - *Oiticella brevis.*
Paineira - *Dysdaemonia boreas.*
Palisade tree - *Arsenura armida* ssp.
Psekamafoura - *Epiphora mythimnia.*
Samalu hedges - *Attacus atlas, A. mcmulleni.*
Trees; deciduous - *Lobobunaea phaedusa.*
Vines - *Loepa katinka* ssp.
Yellow creeper - *Imbrasia rectilineata.*

SUMMARY

This catalog lists some 505 saturniid species and subspecies feeding on 1,060 plant species in 503 genera in 139 families. *Quercus* (FAGACEAE), with 66 species recorded in this catalog, comprises the largest individual number of foodplants in one genus consumed by saturniid larvae. *Pinus* (PINACEAE), with 28 recorded species is the next largest genus. However, the family FABACEAE is represented by three subfamilies, 69 genera and 126 species.

That these three plant families contain the largest number of individual foodplant species consumed by saturniid larvae is not surprising. Although diverse, all these plant groups are cosmopolitan in distribution and represent very ancient orders. As foodplants, they were probably well established before saturniids evolved and, thus, available as larval food sources.

As a group, saturniid larvae consume a phenomenal variety of vegetation, but only species of Spermatophyta—Gymnospermae (conifers) and Angiospermae (flowering plants). Conspicuously absent are records of saturniid larvae feeding on Pteridophyta (ferns and fern allies). Perhaps as research on saturniid foodplants progresses, a pteridophyte may one day be recorded as a saturniid foodplant.

The overwhelming majority of saturniid species listed are from the New World, the Nearctic and Neotropical regions. But this is only because there would appear to be an exceptional richness of saturniid species in these regions. Certainly the saturniid fauna of the western Palaearctic region, Europe, has been well studied by generations of researchers curious about saturniid biology. Far less is known about the biology of many saturniids in the Ethiopian and Indo-Australian regions. Although these areas have been fairly intensively studied and Barlow (1982), Bouvier (1936), Cotes (1891-1893), Dickerson (1944, 1945), Gosse (1879), Holloway (1987), Inoue (1982), Oberprieler (1987), Peigler (1989), Pinhey (1956, 1972, & 1975), Rougeot (1962, 1971), and others have contributed enormous amounts to our understanding of saturniid biology, there are still large gaps in the knowledge.

Very little has been reported from the eastern Palaearctic region, that vast area of the Union of Soviet Socialist Republics. Much of the scant information that is available is generally published in Russian and, for the most part, linguistically inaccessible to many researchers. Perhaps with Glasnost will come accessibility of the natural sciences also.

On a global basis, there are somewhere in the neighborhood of 1,100 to 1,200 named saturniid species and subspecies, depending on the author and whether subspecies are recognized. Represented in this catalog are slightly less than half the described saturniid species. More than anything else, this catalog is a clear and unambiguous statement of the unknown, that is, the remaining 600+ species. Obviously the field of saturniid biology is wide open. With more than 600 species known to science only through dead adults, there is much to learn. It is hoped this work will provide an impetus for research on these remaining species.

REFERENCES CONSULTED

Ainslie, C. N. 1910. The New Mexico range caterpillar. U. S. Dept. Agr. Bur. Entomol. Bull. 85: 59-96.

Albaugh, D. 1986. Foodplants of thirty-six worldwide Saturniidae. *Youth Entomol. Soc.* 39-43.

Aldrich, J. M. 1912. Larvae of a Saturniid moth used as food by California Indians. J. N.Y. Entomol. Soc. 20: 28-31.

_____. 1921. *Coloradia pandora* Blake, a moth of which the caterpillar is used as food by Mono Lake Indians. Ann. Entomol. Soc. Amer. 14: 36-38.

Allen, M. G. 1981. The saturniid moths of Borneo with special reference to Brunei. Brunei Mus. J. 5: 100-126.

André, E. 1908. Élevage des vers à soie sauvages. G. Ficker, Paris. 252 pp.

Arora, G. S. & I. J. Gupta. 1979. Taxonomic studies on some of the Indian non-mulberry silkmoths (Lepidoptera: Saturniidae: Saturniinae). Mem. Zool. Surv. India (Calcutta). 16. 63 pp.

Bailey, L. H. & E. Z. Bailey (comp.). 1976. Hortus third. Macmillan, New York. 1290 pp.

Baker, W. L. 1972. Eastern forest insects. U.S. Dept. Agr. For. Serv. Misc. Publ. 1175. 642 pp.

Barkley, T. M. (ed.). 1977. Atlas of the flora of the Great Plains. Iowa State Univ. Press., Ames. 600 pp.

Barlow, H. S. 1982. An Introduction to the moths of Southeast Asia. Malayan Nature Soc., Kuala Lumpur. 305 pp.

Bauer, D. L. 1948. Methods for collecting buck-moths (*Hemileuca*, Saturniidae). Lep. News 2: 81-82.

Benjamin, F. H. 1934. Notes on the genus *Rothschildia* in the United States. Pan-Pac. Entomol. 10: 12-16.

Benson, L. & R. A. Darrow. 1981. Trees and shrubs of the Southwestern deserts. Univ. of Arizona Press, Tucson. 416 pp.

Beutelspacher B., C. R. 1978. Familias Sphingidae y Saturniidae (Lepidoptera) de Las Minas, Veracruz, México. An. Inst. Biol. Univ. Nal. Autón. México 49, Ser. Zool. (1): 219-229.

Blanchard, A. 1964. The food plants of *Syssphinx heiligbrodti* (Saturniidae) in Texas. J. Lepid. Soc. 18: 42.

Bouvier, E.-L. 1936. Étude des Saturnioïdes normaux. Famille des Saturniidés. Mém. Mus. Natl. Hist. Nat. (Paris). 3: 1-350.

Brodie, W. 1882. Foodplants of *Platysamia cecropia*; *in* H. Edwards, Papilio: The organ of the New York Entomological Club, New York. 2: 32-33.

Brown, D. E. (ed.). 1982. Biotic communities of the American Southwest - United States and Mexico. Desert Plants, Univ. of Arizona Press. 342 pp.

Brown, L. N. 1972. The silkmoths of Florida. Fla Nat. 45: 40-43.

Carr, T. W. 1984. Hybridization between *Callosamia* and *Hyalophora* (Saturniidae). J. Lepid. Soc. 38: 261-267.

Carter, J. L. 1988. Trees and shrubs of Colorado. Johnson Books, Boulder, Colorado. 165 pp.

Collins, M. M. & R. D. Weast. 1961. Wild silk moths of the United States. Collins Radio Co., Cedar Rapids, Iowa. 138 pp.

Collins, M. M. 1973. Notes on the taxonomic status of *Hyalophora columbia* (Saturniidae). J. Lepid. Soc. 27: 225-235.

_____. 1984. Genetics and ecology of a hybrid zone in *Hyalophora* (Lepidoptera: Saturniidae). Univ. Calif.Press., Berkeley. 93 pp.

Comstock, J. A. 1931. Studies in Pacific Coast Lepidoptera. Bull. So. Calif. Acad. Sci. 29: 135-142.

_____. & C. M. Dammers. 1937. Notes on the early stages of three California moths. Bull. So. Calif. Acad. Sci. 36: 68-78.

_____. 1939. Studies in the metamorphoses of six California moths. Bull. S. California Acad. Sci. 37: 105-28.

Coombes, A. J. 1987. Dictionary of plant names. Timber Press, Portland, Oregon. 207 pp.

Correll, D. S. & M. C. Johnston. 1970. Manual of the vascular plants of Texas. Texas Research Foundation, Renner. 1881 pp.

Cotes, E. C. 1891-1893. The wild silk insects of India. Indian Mus. Notes 2: 69-89.

Coyle, J. & N. C. Roberts. 1975. A field guide to the common and interesting plants of Baja California. Natural History Publishing Co., La Jolla, California. 206 pp.

Crotch, W. J. B. 1956. A silkmoth rearer's handbook. Amat. Entomol. Soc. 12. 165 pp. [1969 reprint].

Cryan, J. F. & R. Dirig. 1977. The moths of autumn. Pine Bush Historic Preservation Project., Inc. Occ. Pap. No. 1, Albany, New York. 16 pp.

Dammerman, K. W. 1929. The agricultural zoology of the Malay Archipelago: The animals injurious and beneficial to agriculture, horticulture and forestry in the Malay Peninsula, the Dutch East Indies and the Philippines. J.H. de Bussy, Amsterdam. 473 pp.

Deodikar, G. B., K. K. Kshirsagar & I. A. Kamate. 1969. Chromosome number in *Actias selene* Hübner-A wild silkworm with reliable cocoons. Indian J. Gen. Plant Breeding 29: 126-130.

Dias, M. M. 1978. Morfolofia <sic> e biologia de *Citheronia laocoon* (Cramer, 1777) (Lepidoptera, Adelocephalidae). Rev. Bras. Entomol. 22: 167-197.

Dickerson, C. G. C. 1944. Recently-observed food-plants of some Cape Lepidopterous larvae. J. Entomol. Soc. South. Africa 7: 96-98.

_____. 1945. Recently observed food-plants of some Cape lepidopterous larvae (2nd series). J. Entomol. Soc. South. Africa 8: 151-153.

Dickins, M. 1973. The world of moths. Macmillan, New York. 128 pp.

Dominick, R. B. & C. R. Edwards, 1971. Flight pattern of the male of *Anisota virginiensis* (Citheroniidae). J. Lepid. Soc. 25: 84-85.

Donahue, J. P. 1979. Strategies for survival, the cause of a caterpillar. Tera 17: 3-9.

Dupont, F., G. J. Scheepmaker. [1936]. Uit Java's Vlinderleven. Visser & Co., Batavia. 216 pp.

Edwards, H. 1880. Transformations of some species not hitherto recorded. Pacific Coast Lepidoptera, (New York). No. 27: 2-3.

Eliovson, S. 1956. Flowering shrubs and trees for South African gardens. Howard Timmins, Cape Town. 143 pp.

Eliot, I. A. & C. G. Soule. 1902. Caterpillars and their moths. Century Co., New York. 302 pp.

English, L. L. 1958. Illinois trees and shrubs: Their insect enemies. Ill. Nat. Hist. Survey, Urbana. Circ. 47. 92 pp.

Elmore, E. H. & J. R. Janish. 1976. Shrubs and trees of the Southwest Uplands. Southwestern Parks and Monuments Association, Globe, Arizona. 214 pp.

Ely, R. 1954. Concerning *Hemileuca maia* in Wisconsin. J. Lepid. Soc. 8: 29.

Eyer, J. R. & C. H. Menke. 1914. *Adelocephala bisecta* (Lepidoptera, Family Ceratocampidae). Entomol. News 25: 156-157.

Evans, D. 1958. Two-year life cycle of *Pseudohazis eglanterina* (Boisduval) (Lepidoptera: Saturniidae). Can. Ent. 110: 125-127.

Fauvel, A.-A. 1895. Les séricigènes sauvages de vla Chine. E. Leroux, Paris. 153 pp.

Ferguson, D. C. 1971-1972. Bombycoidea (in part), Saturniidae; *in* R. B. Dominick et al., The moths of America north of Mexico, fasc. 20.2A & 20.2B: 1-275. Wedge Entomol. Foundation & E. W. Classey, London.

_____. & V. A. Brou. 1981. A new species of *Automeris* Hübner (Saturniidae) from the Mississippi River Delta. J. Lepid. Soc. 35: 101-105.

Fernald, M. L. 1970. Gray's manual of botany (8th ed.). D. Van Norstrand, (New York). 1632 pp.

Forbes, W. T. M. 1923. The Lepidoptera of New York and neighboring States. Mem. Agric. Exp. Sta., Cornell Univ., Ithaca. 68: 659-675.

Fruhstorfer, H. 1904. Mitteilung ueber *Attacus atlas* L. und Verwandte. Entomol. Meddel. 2: 283-290.

Furniss, R. L. & V. M. Carolin. 1977. Western forest insects. U. S. Dept. Agr. For. Serv. Misc. Publ. 1339. 654 pp.

Gage, E. V. 1976. Some preliminary notes about the immature stages of *Eacles oslari* (Citheronidae [sic]). J. Res. Lepid. 15: 175-176.

Galtsoff, P. S., F. E. Lutz, P. S. Welch, & J. G. Needham. 1937. Culture methods for invertebrate animals. Comstock Publ. Co., Ithaca. 590 pp.

Gara, R. I. & G. Onore. 1989. Entomologia Foresta. Proyecto Dinaf, Aid Quito, Ecuador. 267 pp.

Gardiner, B. O. C. 1965. The rearing of *Dirphiopsis eumedide* (Saturniidae). J. Res. Lepid. 4: 287-292.

_____. 1966. Notes on *Eacles penelope* (Saturniidae). J. Res. Lepid. 5: 177-180.

_____. 1974. The early stages of various species of the genus *Dirphia* (Saturniidae). J. Res. Lepid. 13: 101-114.

_____. 1976. The early stages of *Leucanella memusae* ssp. *gardineri* Lemaire (Saturniidae). J. Res. Lepid. 15: 201-205.

_____. 1982. A silkmoth rearer's handbook (3rd ed.). Amat. Entomol. Soc. 12. 255 pp.

Gosse, P. H. 1879. *Attacus atlas* : a life-history. Entomol. 12: 25-41, 67-75.

Gould, F. W. & R. Morman. 1981. The grasses of Baja California, Mexico. San Diego Soc. Nat. Hist., California. Mem. 12. 140 pp.

Griveaud, P. 1961. Faune de Madagascar 14. Insectes, Lépidoptères Eupterotidae et Attacidae. Publ. Inst. Sci. Madagascar. 64 pp.

Gunn, C. G. 1983. A nomenclator of Legume (Fabaceae) genera. U. S. Dept. Agric. Tech. Bull. 1680. 224 pp.

Heitzman, R. 1961. The life history of *Adelocephala quadrilineata* (Saturniidae). J. Lepid. Soc. 15: 223-234.

Hepburn, G. A., H. E. Prinsloo, & J. Loedolf. 1966. *Lobobunaea epithyrena* M. & W. (Order Lepidoptera, Family Saturniidae). A potential pest of exotic plantations. For. South. Africa 6: 65-78.

Heppner, J. B., H. Y. Wang, & Y.-C. Chang. 1988. Larval morphology of Taiwan Saturniidae (Lepidoptera): *Actias selene ningpoana*. J. Taiwan Mus. 41: 107-114.

_____, H. Y. Wang, & Y. C. Sen. 1989. Larval morphology of Taiwan Saturniidae (Lepidoptera): *Attacus atlas* (Linnaeus). J. Taiwan Mus. 42: 89-97.

Hill, D. S. & W. W. K. Cheung. 1978. Hong Kong insects. D.R. Rick Gov't. Printer, Hong Kong. 128 pp.

Hitchcock, A. S. 1950. Manual of grasses of the United States (2nd ed.). U. S. Dept. Agr. Misc. Publ. 200. 1051 pp.

Hogue, C. L., F. P. Sala, N. McFarland, & C. Henne. 1965. Systematics and life history of *Saturnia (Calosaturnia) albofasciata* in California (Saturniidae). J. Res. Lepid. 4: 173-184.

Holland, W. J. 1903. The Moth Book. Doubleday, Page and Co., New York. 479 pp.

Holloway, J. D. 1987. The moths of Borneo, part 3. Lasiocampidae, Eupterotidae, Bombycoidea, Brahmaeidae, Saturniidae, Sphingidae. Southdene Sdn. Bhd., Kuala Lumpur, Malaysia. 199 pp.

Inoue, H. 1982. Moths of Japan. Kodansha, Tokyo. 2 vols.

Janzen, D. H. 1982. Guia para identificacion de Mariposas nocturnas de la familia Saturniidae del Parque Nacional Santa Rosa, Guanacaste, Costa Rica. Brenesia 19/20: 255-299.

_____. 1984a. Two ways to be a tropical big moth: Santa Rosa saturniids and sphingids. Oxford Surv. Evol. Biol. 1: 85-140.

_____. 1984b. Natural History of *Hylesia lineata* (Saturniidae: Hemileucinae) in Santa Rosa National Park, Costa Rica. J. Kansas Entomol. Soc. 57: 490-514.

_____. 1985. A host plant is more than its chemistry. Illinois Nat. Hist Survey Bull. 33: Art. 3, 141-174.

Jones, F. M. 1909. Additional notes on *Callosamia carolina*. Entomol. News 20: 49-51.

Jones, J. R. J. 1851. An annotated check list of the Macrolepidoptera of British Columbia. Entomol. Soc. Brit. Columbia Occ. Pap. 1: 148 pp.

Jurriaanse, J. H. & J. Lindemans. 1920. Indo-Australische Attacus-vormen. Tijds. Entomol. 63: 87-95.

Kearney, T. H. & R. H. Peebles. 1942. Flowering plants and ferns of Arizona. U. S. Dept. Agr. Misc. Publ. 423. 1069 pp.

Kelly, G. W. 1976. Trees for the Rocky Mountains. Rocky Mountain Horticultural Publ. Co., World Press. 119 pp.

_____. 1979. Shrubs for the Rocky Mountains. Rocky Mountain Horticultural Publ. Co., Cortez, Colorado. 171 pp.

Kendall, R. O. & R. S. Peigler. 1981. *Hemileuca grotei* (Saturniidae): Its morphology, natural history, spatial, and temporal distribution. J. Lepid. Soc. 35: 41-50.

Kimball, C. P. 1965. The Lepidoptera of Florida. Arthropods of Florida and Neighboring Land Areas. 1: 363 pp.

Köhalmi, L. & P. Moens. 1975. Evidence for the existence of an intergrade population between *Hyalophora gloveri nokomis* and *H. columbia* in Northwestern Ontario (Lepidoptera: Saturniidae). Can. Ent. 107: 793-799.

Kunzé, R. E. 1900. Notes on ova and larva of *Hyperchiria pamina*. J. New York Entomol. Soc. 8: 201-205.

Lampe, R. E. J. & W. A. Nässig. 1989. Neue Erkenntnisse über die Gattung *Lemaireia*: 1. Beschreibung der Präimaginalstadien von *L. luteopeplus aureopeplus* Nässig & Holloway (Lepidoptera, Saturniidae). Nachr. Entomol. Ver. Apollo, Frankfurt, N.F. 10: 225-238.

Laporte, J. L. 1830. Observations pour servir à l'historie de quelques insectes, et description d' une nouvelle espèce de Coliade. Act. Soc. Linn. Bordeaux 4: 141-156.

_____. 1906. Descriptions of two lepidopterous larva. J. N.Y. Entomol. Soc. 14: 131-134.

Lamb, S. H. 1975. Woody plants of the Southwest. Sunstone Press, Santa Fe, New Mexico. 177 pp.

Leeuw, I. 1974. A further note on the acceptability of an alternate foodplant for *Hemileuca maia* (Drury) (Saturniidae). J. Lepid. Soc. 28: 301.

Lemaire, C. 1971-1974. Révision du genre *Automeris* Hübner et des genres voisins biogéographie, éthologie, morphologie, taxonomie (Lep. Attacidae). Mém. Mus. Natl. Hist. Nat., Sér. A, Zool. (Paris). 68: 1-232, 79: 233-422, 92: 423-576.

_____. 1978. The Attacidae of America. Attacinae. Édition C. Lemaire, Neuilly-sur-Seine, France. 238 pp.

_____. 1979a. A new subspecies of *Hemileuca maia* from central Texas (Attacidae, Hemileucinae). J. Res. Lepid. 18: 212-219.

_____. 1979b. A new species of *Automeris cecrops* (Attacidae: Hemileucinae). J. Res. Lepid. 18: 236-240.

_____. 1980. The Attacidae of America. Arsenurinae. Édition C. Lemaire, Neuilly-sur-Seine, France. 199 pp.

_____. 1988. The Saturniidae of America, Ceratocampinae. Museo Nacional de Costa Rica, San José. 469 pp.

_____. & K. L. Wolfe. 1988. Three new species of *Paradirphia* (Saturniidae: Hemileucinae) from Mexico and Central America with notes on the immature stages. J. Res. Lepid. 27: 197-212.

Little, E. L. 1950. Southwestern trees: A guide to the native species of New Mexico and Arizona. U. S. Dept. Agr. For. Ser. Handb. 9. 109 pp.

_____. 1979. Checklist of United States trees (native and naturalized). U. S. Dept. Agr. For. Ser. Handb. 541. 375 pp.

Lowe, C. H. 1964. Arizona's natural environment: Landscapes and Habitats. Univ. of Arizona Press, Tucson. 136 pp.

Mason, C. T. & P. B. Mason. 1987. A handbook of Mexican roadside flora. Univ. of Arizona Press, Tucson. 380 pp.

McGugan, B. M., W. H. Haliburton, & J. E. MacDonald. 1952. Forest Insect Survey. Ann. Rept. For. Insect and Disease Surv., Canada Dept. Agric. for 1951, pp. 39-65.

_____. (comp.). 1958. Forest Lepidoptera of Canada recorded by the Forest Insect Survey, I, Papilionidae to Arctiidae. Publ. 1034, For. Biol. Div., Canada Dept. Agric. 76 pp.

McFarland, N. 1974. Notes on three species of *Hemileuca* (Saturniidae) from eastern Oregon and California. J. Lepid. Soc. 28: 136-141.

_____. 1975. Larval foodplant records for 106 species of North American moths. J. Lepid. Soc. 29: 112-125.

Mell, R. 1950. Aus der Biologie der chineschen *Actias* Leach. Entomol. Z. (Stuttgart) 60: 41-45, 53-65.

Michener, C. D. 1950. A northern subspecies of *Eacles imperialis* (Lepidoptera, Saturniidae). J. Kansas Entomol. Soc. 23: 17-21.

Miller, H. A. & S. H. Lamb. 1985. Oaks of North America. Naturegraph Publishing Co., Happy Camp, California. 327 pp.

Moore, F. 1859. Synopsis of the known Asiatic species of silk-producing moths, with descriptions of some new species from India. Proc. Zool. Soc. London 27: 237-270.

_____. 1862. On the Asiatic silk-producing moths. Trans. Entomol. Soc. London 1: 313-322.

Morris, J. G. 1862. Synopsis of the described Lepidoptera of North America. Part I.—Diurnal and Crepuscular Lepidoptera. Smithsonian Misc. Coll., Smithsonian Inst. 358 pp.

Moucha, J. 1966. Beautiful moths. Spring Books, London. 139 pp.

Munro, H. K. 1946. A lengthened pupal period for a moth. J. Entomol. Soc. South. Africa 9: 3.

Munz, P. A. 1968. A California flora. Univ. of California Press, Berkeley. 224 pp.

Nardelli, U. 1986. Zur Kenntnis der Präimaginalstadien und zur Zucht von *Caligula thibeta* Westwood, 1853 (Lepidoptera, Saturniidae). Nachr. Entomol. Ver. Apollo, Frankfurt, N. F. 7: 11-23.

Nässig, W. A. & R. S. Peigler. 1984. The life-history of *Actias maenas* (Saturniidae). J. Lepid. Soc. 38: 114-123.

Neck, R. W. 1976. Lepidopteran foodplant records from Texas. J. Res. Lepid. 15: 75-82.

Oberprieler, R. G. 1986. *Pinus radiata* D. Don: A new foodplant record for *Imbrasia* (*Gonimbrasia*) *belina* (Westwood) (Lepidoptera: Saturniidae). J. Entomol. Soc. South. Africa. 49: 168-170.

_____. 1987. Emperor Moths (Saturniidae) as Botanists - The value of foodplant data in classification and conservation. *in* Proc. 1st. Lepid. Cons. Symp., Roodepoort. Lepid. Soc. of South. Africa. pp. 18-21.

Packard, A. S. 1893. The life histories of certain moths of the families Ceratocampidae, Hemileucidae, etc., with notes on the armature of the larvae. Proc. Amer. Philosophical Soc. 31(141): 139-192.

_____. 1905. Monograph of the Bombycine moths of North America, including their transformations and origin of the larval markings and armature. Part 2, Family Ceratocampidae, subfamily Ceratocampinae. Mem. Natl. Acad. Sci. 9: 1-272 pp.

_____. 1914. Monograph of the Bombycine moths of North America, part III, (ed. by T. D. A. Cockerell). Mem. Natl. Acad. Sci. 12: 1-276, + 503-516.

Patraw, P. M. & J. R. Janish. 1977. Flowers of the Southwest mesas. Southwest Parks and Monuments Association, Globe, Arizona. 112 pp.

Patterson, J. E. 1929. The pandora moth, a periodic pest of western pine forests. U. S. Dept. Agr. Tech. Bull. 137: 19 pp.

Peigler, R. S. 1976. Observations on host plant relationships and larval nutrition in *Callosamia* (Saturniidae). J. Lepid. Soc. 30: 184-187.

_____. 1977. Hybridization of *Callosamia* (Saturniidae). J. Lepid. Soc. 31: 23-34.

_____. 1978. Hybrids between *Callosamia* and *Samia* (Saturniidae). J. Lepid. Soc. 32: 191-197.

_____. 1981. Demonstration of reproductive isolating mechanisms in *Callosamia* (Saturniidae) by artificial hybridization. J. Res. Lepid. 19: 72-81.

_____. 1985. Description, distribution, and biology of a new *Hemileuca* (Lepidoptera: Saturniidae) from western Texas and southern New Mexico. Nachr. Entomol. Ver. Apollo, Frankfurt, N.F. 2: 113-124.

_____. 1988. Hostplants of *Callosamia* (Saturniidae) and *Epimecis* (Geometridae) with special reference to the magnolia, laurel, and tea families. Nachr. Entomol. Ver. Apollo, Frankfurt, N.F. 9: 91-100.

_____. 1989. A revision of the Indo-Australian genus *Attacus*. Lepidoptera Research Foundation, Beverly Hills, California. 167 pp.

_____. & S. E. Stone. 1989. Taxonomic and biological notes on the *Hemileuca maia* complex (Lepidoptera: Saturniidae) with description of a new species from Texas and New Mexico. Lepid. Soc. Japan 40: 149-165.

_____. & B. D. Williams. 1984. Two interesting artificial hybrid crosses in the genera *Hemileuca* and *Anisota* (Saturniidae). J. Lepid. Soc. 38: 51-56.

Perry, B. 1987. Trees and shrubs for dry California landscapes: Plants for water conservation. Land Design Publishing. San Dimas, California. 184 pp.

Pinhey, E. 1956. The Emperor moths of eastern Africa. J. E. Afr. Nat. Hist. Soc. 23: 1-63.

_____. 1968. The genus *Gynanisa* Walker (Lepidoptera: Saturniidae). J. Entomol. Soc. South. Africa 31: 123-132.

_____. 1972. Emperor moths of South and South-Central Africa. C. Struik (PTY) Ltd., Cape Town. 150 pp.

_____. 1975. Moths of southern Africa. Tafelberg Publ., Ltd., Cape Town. 273 pp.

Platt, A. P. & T. F. Williams. 1982. Differential growth and utilization of three foodplants by first instar larvae of *Citheronia regalis* (Saturniidae). J. Lepid. Soc. 36: 76-82.

Platt, E. E. 1921. List of foodplants of some South African lepidopterous larvae. South African J. Nat. Hist. 3: 99-138.

Poujade, G.-A. 1880. Observations sur les métamorphoses de l'Attacus Atlas. Ann. Soc. Entomol. France 10: 183-188.

Preston, R. J.. 1968. Rocky Mountain trees (3rd ed). Dover Publishing Co., New York. 285 pp.

Priddle, T. R. 1966. Techniques for reducing mortality when rearing larvae of cecropia moth (Saturniidae). J. Lepid. Soc. 119-121.

Rama Rao, R. S. M. 1914. Flowering plants of Travancore. Government Press, Trivandrum. 495 pp.

Reiff, W. 1910. Notes on *Hemileuca lucina* Hy. Edw. Psyche 17: 29-32.

Riley, C. V. 1889. Larva of *Hyperchiria io* on Saw Palmetto in Florida. Insect Life 1: 217.

Riotte, J. C. E. & R. S. Peigler. 1981. A revision of the American genus *Anisota* (Saturniidae). J. Res. Lepid. 19: 101-180.

Rougeot, P.-C. 1962. Attacidés (=Saturniidés). *In*: Initiations et études of Africaines. 14 Les Lépidoptères de D'Afrique Noire Occidentale. Bull. Inst. Franc. Afr. Noire 4: 205 pp.

_____. 1971. Les Bombycoïdes (Lepidoptera-Bombycoïdea) de l'Europe et du Bassin Méditerranéen. Masson et Cie (Paris). 159 pp.

_____. & R. Viette. 1983. Die Nachtfalter Europas und Nordafrikas, I. Schwärmer und Spinner (1. Teil). E. Bauer, Keltern, West Germany. 281 pp.

Sala, F. P. & C. L. Hogue. 1958. Descriptions of the early stages and male imago and notes on the life history of *Saturnia walterorum* (Saturniidae). Lep. News 12: 17-25.

Saunders, W. 1909. Insects injurious to fruits (2nd ed). J. B. Lippincott, London. 436 pp.

Scarbrough, A. G., G. P. Waldbauer, & J. G. Sternburg. 1974. Feeding and survival of *cecropia* (Saturniidae) larvae on various plant species. J. Lepid. Soc. 28: 212-219.

Schaus, W. 1884. Early Stages of Mexican Lepidoptera. Papilio 14: 100-103.

Schopmeyer, C. S. (tech. coord.). 1974. Seeds of woody plants in the United States. U. S. Dept. Agr., Agr. Hndbk. No. 450. 883 pp.

Seitz, A. 1926. Saturniidae. *In* A. Seitz, Die Gross-Schmetterlinge der Erde 10: 497-520. Alfred Kernen, Stuttgart.

Shelton, N. 1972. Saguaro National Monument, Arizona. U. S. Dep. Inter., Nat. Park Ser., Washington, D. C. 98 pp.

Silbermann, H. [1897]. Die Seide, ihre Geschichte, Gewinning und Verarbeitung 1. Leipzig.

Skinner, H. 1908. Note by Dr. Henry Skinner. Can. Entomol. 40: 354.

Smith, J. E. 1797. The natural history of the rarer lepidopterous insects of Georgia, collected from the observations of John Abbot, with the plants on which they feed. 2 vols. Edwards, London.

Smith, M. J. 1974. Life history notes on some *Hemileuca* species (Saturniidae). J. Lepid. Soc. 28: 272-276.

_____. 1988. The Hemileucinae of Utah. Utahensis 8(3): 26-33.

Smith, S. I. 1865. Description of a species of *Samia*, supposed to be new from Norway, Me. Proc. Boston Soc. Nat. Hist. 9: 342-345.

SsuMa, J. 1972. Moth with snake-head pattern. Harvest, Farm Mag. 22: 46-47.

Standley, P. C. 1920-1926. Trees and shrubs of Mexico. Contrib. U. S. Nat. Herbarium 23, pts. 1-5. 1721 pp.

_____. 1930. Flora of Yucatan. Field Mus. Natl. Hist. Publ. 279, 3(3). 492 pp.

Sternburg, J. G. & G. P. Waldbauer. 1981. Distribution of cecropia moth (Saturniidae) in central Illinois: A study in urban ecology. J. Lepid. Soc. 35: 304-320.

Steyermark, J. A. 1981. Flora of Missouri. Iowa State Univ. Press, Ames, Iowa. 1728 pp.

Stone, S. E., D. E. Swift, & R. S. Peigler. 1988. The life history of *Hemileuca magnifica* (Saturniidae) with notes on *Hemileuca hera marcata*. J. Res. Lepid. 26: 225-235.

_____. & M. J. Smith. 1990. Buckmoths (Lepidoptera: Saturniidae: *Hemileuca*) in relation to Southwestern vegetation and foodplants. Desert Plants 10(1): 13-18, 23-30.

Sweadner, W. R. 1937. Hybridization and the phylogeny of the genus *Platysamia*. Ann. Carnegie Mus. 25: 163-242.

Taylor, J. S. 1947. Some observations on "assembling". J. Entomol. Soc. South. Africa 10: 179-181.

_____. 1957. Notes on Lepidoptera in the Eastern Cape Province (Part IV). J. Entomol. Soc. South. Africa 20: 315-332.

Tietz, H. M. 1972. An index to the described life histories, early stages and hosts of the Macrolepidoptera of the continental United States and Canada. A. C., Allyn, Sarasota, Florida. 1. 536 pp.

Tilden, J. W. 1945. Notes on some moths of the family Saturniidae. Pan-Pac. Entomol. 21: 32-33.

Townsend, C. H. T. 1893. Notes on *Hemileuca*. W. Amer. Sci. 8: 51-53.

Tuskes, P. M. 1974. The distribution and larval foodplant relationships of *Saturnia walterorum* (Saturniidae). J. Lepid. Soc. 28: 172-174.

_____. 1976. A key to the last instar larvae of west coast Saturniidae. J. Lepid. Soc. 30: 272-276.

_____. 1978. A new species of *Hemileuca* from the southwestern United States (Saturniidae). J. Lepid. Soc. 30: 272-276.

_____. 1984. The biology and distribution of California Hemileucinae (Saturniidae). J. Lepid. Soc. 38: 281-309.

_____. 1985a. The biology and immature stages of *Sphingicampa albolineata* and *S. montana* in Arizona (Saturniidae). J. Lepid. Soc. 39: 85-94.

_____. 1985b. The biology and immature stages of *Automeris randa* and *Automeris iris hesselorum*. J. Lepid. Soc. 39: 163-170.

_____. 1986a. Biology and immature stages of *Hemileuca diana* and *H. grotei* (Saturniidae). J. Lepid. Soc. 40: 27-35.

_____. 1986b. Biology and immature stages of *Citheronia splendens sinaloensis* and *Eacles oslari* in Arizona (Saturniidae). J. Lepid. Soc. 40: 264-270.

_____. & M. M. Collins. 1981. Hybridization of *Saturnia mendocino* and *S. walterorum*, and phylogenetic notes on *Saturnia* and *Agapema* (Saturniidae). J. Lepid. Soc. 35: 1-21.

_____. & M. J. Smith. 1984. The life history and immature stages of *Agapema homogena* (Saturniidae). J. Lepid. Soc. 38: 134-137.

_____. 1988. The life history of *Automeris zephyria* (Saturniidae). J. Res. Lepid. 27: 192-196.

Tuttle, J. P. 1985. Maintaining species integrity between sympatric populations of *Hyalophora cecropia* and *Hyalophora columbia* (Saturniidae) in central Michigan. J. Lepid. Soc. 39: 65-84.

Van den Berg, M. A. & M. M. Van den Berg. 1973. The food assimilation and duration of larval instars of three saturniid forest pests. J. Entomol. Soc. South. Africa 36: 165-173.

Van den Berg, M. A. 1974. Flight activity, dispersion and longevity of adults of *Nudaurelia cytherea clarki* Geertsema (Lepidoptera: Saturniidae). J. Entomol. Soc. of South. Africa 37: 67-71.

Van Dersal, W. R. 1938. Native woody plants of the United States: Their erosion-control and wildlife value. U. S. Dept. Agr. Misc. Publ. 303. 362 pp.

Van Rensselaer, M. & H. E. McMinn. 1942. *Ceanothus*. Gillick Press, Berkeley, California. 308 pp.

Vasquez, L. 1944. *Citheronia splendens queretana* subspec. nov. An. Inst. Biol., México. 15: 235-236.

Ver Huell, Q. M. R. 1836. de rups van den *Bombax atlas*. Tijds. Nat. Gesch. Physiol. 3: 29-30.

Villiard, P. 1969. Moths and how to rear them. Funk & Wagnalls, New York. 242 pp.

Vines, R. A. 1976. Trees, shrubs and woody vines of the Southwest. Univ. of Texas Press, Austin. 1104 pp.

Voelschow, A. 1902. Die Zucht der Seidenspinner. Ausführliche Beschreibung sämtlicher in Europa eingefuhrten seideerzeugenden Schmetterlinge und ihrer Zchwerin. Mecklengurg. 83 pp.

Wagner, W. H. & M. R. Mayfield. 1980. Foodplants and cocoon construction in *Antheraea polyphemus* (Lepidoptera: Saturniidae) in southern Michigan. Great Lakes Entomol. 13: 131-138.

_____, M. K. Hansen, & M. R. Mayfield. 1981. True and false foodplants of *Callosamia promethea* (Lepidoptera: Saturniidae) in southern Michigan. Great Lakes Entomol. 14: 159-165.

Wailly, A. 1879. Notes on certain silk-producing bombyces. Entomol. 12: 8-11.

Waldbauer, G. P. & J. G. Sternburg. 1973. Polymorphic termination of diapause by cecropia: genetic and geographical aspects. Biol. Bull. 145: 627-641.

Wang, H. Y. 1988. Saturniid moths of Taiwan. Taiwan Museum, Taipei. 69 pp.

Watt, G. 1908. The commercial products of India being an abridgment of "The dictionary of the economic products of India". John Murray, London. 1189 pp.

Weast, R. D. 1962. Two new foodplants of southwestern Saturniidae. J. Lepid. Soc. 16: 61-62.

Welsh, S. L., N. D. Atwood, L. C. Higgins, & S. Goodrich. 1987. A Utah flora. Great Basin Nat. Mem. 9. 894 pp.

White, F. 1983. The vegetation of Africa: A descriptive memoir to accompany the Unesco/AETFAT/UNSO vegetation map of Africa. UNESCO, Paris. 356 pp.

White, T. C. R. 1972. The distribution, dispersal and host range of *Antheraea eucalypti* (Lepidoptera: Saturniidae) in New Zealand. Pacific Insects 14: 669-673.

Wiggins, I. L. 1980. Flora of Baja California. Stanford Univ. Press, Stanford, California. 1025 pp.

Wolfe, K. L. 1988. *Hylesia acuta* (Saturniidae) and its aggregate larval and pupal pouch. J. Lepid. Soc. 42: 132-137.

Wright, D. A. 1967. The effects of photoperiod on the initiation of pupal diapause on the wild silkworm, *Actias luna*. J. Lepid. Soc. 21: 255-258.

_____. 1971. Hybrids among species of *Hyalophora*. J. Lepid. Soc. 25: 68-73.

Wyrant, N. D. 1941. An infestation of the pandora moth, *Coloradia pandora* Blake, in lodgepole pine in Colorado. J. Econ. Entomol. 34: 697-702.

Zhu, H. & L. Wang. 1983. Saturniidae, pp. 408-414, *in* Wang, L. & Z. Hongfu. Icon. Hetero. Sinico., Science Press, Beijing. 4: 391-484.

APPENDIX I: VERNACULAR NAMES OF FOODPLANTS

ABIES	Fir
balsamea	Balsam fir
grandis	Grand fir
ACACIA	Acacia\Blackwood\Wattle
angustissima	Prairie acacia
baileyana	Bailey's acacia
collinsii	Bullhorn acacia
cornigera	Bull's horn acacia
cyanophylla	Wattle
cyclopis	Wattle
dealbata	Silver wattle
farnesiana	Huisache\Sweet acacia
flexicaulis	Ebony blackbead
greggi	Catclaw acacia
karroo	Karro wattle
mearnsii	Black wattle
mollissima	Black wattle
nilotica	Gum-Arabic wattle
rigidula	Blackbrush wattle
saligna	Weeping wattle
sieberiana	Wattle
ACALYPHA	Three-seeded Mercury
ACER	Maple
circinatum	Vine maple
dasycarpum	Silver maple
monspessulanum	Montpelier maple
negundo	Box elder/Manitoba maple
plantanoides	Norway maple
pseudoplatanus	Sycamore maple
rubrum	Red maple
saccharinum	Silver maple
saccharum	Sugar maple
spicatum	Mountain maple
ADENOSTOMA	Chamise
AESCULUS	Horsechestnut
hippocastanum	European horsechestnut
AGROPYRON	
smithii	Western wheatgrass
AILANTHUS	Chinese Tree of heaven
ALBIZIA	Silktree
saman	Rain tree
ALEURITES	Tungtree
fordii	China wood-oil tree
montana	Mu-oil tree
ALLOPHYLUS	
occidentalis	Palo blanco
ALNUS	Alder
glutinosa	European\Black alder
incana	White alder
japonica	Japanese alder
rubra	Red alder
serrulata	Hazel alder
ALSTONIA	
scholaris	Devil tree
ALTHAEA	
rosa	Hollyhock
AMELANCHIER	Serviceberry
alnifolia	Western serviceberry
canadensis	Thicket serviceberry
florida	Western serviceberry
AMORPHA	False indigo
fruticosa	Bastard indigo
ANACARDIUM	Cashew
occidentale	Cashewnut\Marañón
ANDROMEDA	Rosemary
glaucophylla	Downy bog-rosemary
ovalifolia	Bog-rosemary
ANDROPOGON	Beardgrass
hallii	Sand bluestem
ANNONA	Annona
muricata	Prickly custard apple
purpurea	Black-bead
senegalensis	Wild custard apple
squamosa	Custard apple
ANOGEISSUS	
latifolia	Dhaura
APIUM	
graveolens	Celery
APODYTES	
dimidiata	White pear

ARBUTUS	Madrone
menziesii	Pacific madrone
unedo	Strawberry madrone
ARCTOSTAPHYLOS	Manzanita
alpinus	Black bearberry
glauca	Great-berried manzanita
patula	Greenleaf manzanita
pungens	Pointleaf manzanita
tomentosa	Shaggy-barked manzanita
uva-ursi	Kinnikinnick
ARONIA	Chokeberry
arbutifolia	Red chokeberry
ARTABOTRYS	Tall grape
ARTEMISIA	Sagebrush\Wormwood
californica	California sagebrush
cana	Silver sagebrush
filifolia	Sand sagebrush
tridentata	Big sagebrush
vulgaris	Mugwort
ARTOCARPUS	
integer	Jackfruit
ASTRONIUM	
graveolens	Kulimche
AVERRHOA	
bilimbi	Cucumber tree
carambola	Country gooseberry
AZADIRACHTA	
indica	Persian lilac
BAPTISIA	False Indigo
tinctoria	Wild Indigo
BASSIA	Mohwa tree
BAUHINIA	Kanchan
variegata	Mountain ebony
BEQUAERTIODENDRON	Fluted milkweed
BERBERIS	Barberry
thunbergii	Japanese barberry
vulgaris	Common barberry
BERLINIA	Mnondo
BETULA	Birch
alba	White birch
alleghaniensis	Yellow birch
alnoides	Indian birch
excelsa	Low birch
lenta	Sweet birch
occidentalis	Mountain\Water birch
papyrifera	White\Paper birch
pendula	Silver birch
platyphylla	Japanese white birch
populifolia	Gray birch
pumila	Dwarf birch
BISCHOFIA	
javanica	Toog
BIXA	
orellana	Annatto
BOMBAX	
ceiba	Red silk-cotton tree
heptaphyllum	Semul
malabaricum	Silk-cotton tree
BOUTELOUA	Grama
curtipendula	Side-oats grama
gracilis	Blue grama
hirsuta	Hairy grama
BRACHYSTEGIA	Msasa
BRASSAIA	
actinophylla var. *arboricola*	Dwarf Australian umbrella tree
BROMUS	Bromegrass
BUCHLOE	
dactyloides	Buffalo grass
BURKEA	
africana	Wild Syringa
BURSERA	Bursera
microphylla	Elephant tree
simaruba	Gumbo-limbo
tomentosa	Torchwood
BYRSONIMA	Byrsonima
crassifolia	Nanche
BYTTNERIA	Tezak\Zarza
CALLIANDRA	Powderpuff
CALLUNA	Heather
vulgaris	Scottish heathers
CALYCOPHYLLUM	
candidissimum	Degame
CANANGA	
odorata	Ilang-ilang

CANARIUM	Java almond
CAPSICUM	Pepper
annuum	Cayenne pepper
CAREYA	Kumbi
CARICA	Papaya
papaya	Pawpaw
CARISSA	Karaunda
carandas	Karanda
grandiflora	Natal plum
CARPINUS	Hornbeam\Ironwood
americana	American hornbeam
betulus	European hornbeam
caroliniana	American hornbeam
CARYA	Hickory
alba	Mockernut hickory
cordiformis	Bitternut hickory
glabra	Pignut hickory
illinoensis	Pecan
ovata	Shagbark hickory
CASSIA	Cape Laburnum\Senna
alata	Ringworm cassia
corymbosa	Autumn senna
fistula	Indian laburnum
grandis	Horse cassia
CASTANEA	Chestnut\Chinkapin
crenata	Japanese chestnut
dentata	American chestnut
mollissima	Chinese chestnut
pumila	Allegheny chinquapin
sativa	Spanish\Sweet chestnut
vesca	Spanish chestnut
vulgaris	American chestnut
CATALPA	Catawba\Indian bean
CEANOTHUS	Wild lilac
cordulatus	Mountain white-thorn
cuneatus	Wedgeleaf ceanothus
fendleri	Fendler ceanothus
integerrimus	Deer brush
leucodermis	Chaparral whitethorn
spinosa	Greenbark ceanothus
thyrsiflorus	Blueblossom
velutinus	Snowbrush
CEDRELA	
paniculata	Cedro\Spanish cedar
odorata	West Indian Cedar
CEDRUS	Cedar
CEIBA	
acuminata	Kapok
pentandra	White silk cotton
CELASTRUS	Shrubby bittersweet
paniculatus	Malkangni
CELTIS	Hackberry
laevigata	Texas sugarberry
sinensis	Japanese hackberry
CENTROSEMA	Butterfly pea
pubescens	Kantsin
CEPHALANTHUS	Buttonbush
CEPHALARIA	
procera	Teasel
CERCIDIUM	Paloverde
floridum	Texas paloverde
microphyllum	Little-leaf horsebean
CERCIS	
canadensis	Redbud
siliquastrum	Judas tree
CERCOCARPUS	Mountainmahogany
betuloides	Birchleaf cercocarpus
montanus	True mountainmahogany
CHAMAECYPARIS	False cypress
thyoides	Atlantic white-cedar
CHILOPSIS	
linearis	Desert willow
CHLORIS	Finger grass
virgata	Rhodes grass
CHORISIA	Floss-silk tree
CHLOROPHORA	Muvuli
CHLOROXYLON	Satin-wood
swietenia	East Indian Satin-wood
CHRYSOPHYLLUM	Goldenleaf
CINCHONA	Quinine
CINNAMOMUM	Cinnamon
camphora	Camphor
iners	Wild cinnamon
obtusifolium	Camphor
zeylanicum	Cinnamon

CISSUS	Grape Ivy
CITRUS	Citrus
aurantium	Sour orange
reticulata	Mandarin orange
CLERODENDRUM	Glory-bower
COCHLOSPERMUM	
vitifolium	Rosa amarillo\Yellow rose
COFFEA	Coffee
COLEOGYNE	
ramosissima	Blackbrush
COLEUS	Painted leaves
barbatus	Flame nettle
blumei	Painted nettle
COLOPHOSPERMUM	Mopane
COMBRETUM	Red wings
COMMIPHORA	Paper tree
COMPTONIA	
peregrina	Sweet fern
CONDALIA	Bluewood
ericcides	Javelina bush
obovata	Bluewood
spathulata	Squawbush
CONOCARPUS	Button mangrove
erectus	Buttonwood
CONOPODIUM	
denudatum	Earth-nut
CORDIA	
alliodora	Apacahuite\Trompillo
CORIARIA	
nepalensis	Tanner's tree
CORNUS	Dogwood
controversa	Giant dogwood
florida	Flowering dogwood
racemosa	Gray dogwood
stolonifera	Red osier dogwood
CORYLUS	Hazel
americana	American hazelnut
avellana	European hazel
colurna	Turkish hazel
cornuta	California hazelnut
heterophylla	Hazel
rostrata	Beaked hazelnut
CRATAEGUS	Hawthorn
calpodendron	Pear hawthorn
coccinea	Scarlet hawthorn
crus-galli	Cockspur hawthorn
oxyacantha	English hawthorn
pedicellata	Scarlet hawthorn
tomentosa	Pear hawthorn
CRESCENTIA	Calabash tree
CUPANIA	
guatemalensis	Sacpom\Grande Betty
CUPRESSUS	Cypress
macrocarpa	Monterey Cypress
CURCUMA	Hidden lily
longa	Turmeric
CUSSONIA	
spicata	Cabbage tree
CYCLOBALANOPSIS	
glauca	Japanese evergreen oak
CYDISTA	
heterophylla	Chacnetoloc
CYDONIA	Quince
oblonga	Common quince
CYNODON	Bermuda grass
CYPERUS	Umbrella sedge
esculentus	Nutsedge
DALEA	Indigobush
fremontii	Fremont dalea
DELONIX	Flamboyant
DIALIUM	
guineense	Velvet tamarind
DIERVILLA	Bush honeysuckle
DIOSPYROS	Persimmon
kaki	Japanese persimmon
lycioides	Ebony
mespiliformis	Ebony
virginiana	Common persimmon
DISTICHLIS	
spicata	Seashore saltgrass
DODONAEA	
viscosa	Hopbush
ELAEAGNUS	
angustifolia	Russian olive
argentea	Silverberry

ENTEROLOBIUM	
cyclocarpum	Orejón\Elephant's-ear
EPHEDRA	Mormon-tea
nevadensis	Nevada jointfir
ERICA	Heath
ERIOBOTRYA	
japonica	Loquat
ERIOGONUM	Buckwheat
effusum	Spreading buckwheat
fasciculatum	California buckwheat
fasciculatum var. *polifolium*	Rosemary flat-top eriogonum
wrightii	Wright wildbuckwheat
ERIOSEMA	Blue bush
ERYNGIUM	Eryngo
ERYTHRINA	Kaffir boom\Coral tree
crista-galli	Cockspur coral tree
ERYTHROPHLEUM	
suaveolens	Red-water tree
ERYTHROXYLUM	
havanense	Coca
EUCALYPTUS	Eucalyptus\Gum
baxteri	Gum
blakelyi	Blakeley's red gum
bridgesiana	Apple box
cinerea	Mealy Stringybark
eugenioides	White Stringybark
fastigiata	Brown-barrel
globulus	Tasmanian Blue Gum
gunnii	Cider gum
leucoxylon	White Ironbark
leucoxylon var. *macrocarpa*	Blue Gum
macarthuri	Camden Wollybutt
maculosa	Red-spotted Gum
macrorhyncha	Red Stringybark
maideni	Maiden's Gum
melliodora	Yellow Box
nitens	Gum
odorata	Gum
obliqua	Messmate
paniculata	Gray Ironbark
pauciflora	Cabbage Gum
regnans	Giant Gum
saligna	Sydney Blue Gum
viminalis	Manna Gum
EUCLEA	Ebony
EUGENIA	Stopper\Waterboom
jambolana	Jáman
EUPHORBIA	
longana	Eye-ball tree
EUODIA	
fraxinifolia	Poyam
EXOSTEMA	
mexicanum	Princewood
EYSENHARDTIA	
polystachya	Kidneywood
FAGUS	Beech
crenata	Japanese beech
ferruginea	American beech
grandifolia	American beech
sylvatica	European beech
FALLUGIA	Apache plume
FICUS	Fig
benjamina	Benjamin Tree
cordata	Fig
religiosa	Sacred Fig
retusa	Indian Laurel
FIRMIANA	
simplex	Chinese parasol tree
FORESTIERA	
angustifolia	Narrow-leaf forestiera
FORSYTHIA	Golden-bells
FOUQUIERIA	Ocotillo
FRAGARIA	Strawberry
FRAXINUS	Ash
americana	White ash
berlandierana	Berlandier ash
excelsior	European ash
pennsylvanica	Green ash
FREMONTIA	Flannel bush
GAYLUSSACIA	Huckleberry
dumosa	Dwarf huckleberry
frondosa	Dangleberry
GINKGO	Maidenhair tree

GLEDITSIA	Honeylocust
texana	Texas honeylocust
triacanthos	Honeylocust
GLIRICIDIA	
sepium	Nicaraguan cocoa-shade
GLYCYRRHIZA	American licorice
GMELINA	
arborea	Gomari
GORDONIA	
lasianthus	Loblolly-bay
GOSSYPIUM	Cotton
herbaceum	Levant cotton
hirsutum	Upland Cotton
thurberi	Arizona wild cotton
GOUANIA	
polygama	Jaboncillo
GUAZUMA	
ulmifolia	Guacima\Bay-cedar
GYMNOCLADUS	Coffee tree
canadensis	Coffee tree
dioica	Kentucky coffee tree
HALESIA	Silver-bell
carolina	Wild olive
HAMAMELIS	Witch hazel
HARPEPHYLLUM	
caffrum	Kaffir plum
HELICONIA	Red lobster-claw
HETEROPANAX	
fragrans	Keseru
HEVEA	
brasiliensis	Para rubber tree
HIBISCUS	Mallow
syriacus	Syrian hibiscus
tiliaceus	Sea hibiscus
HILARIA	
cenchroides	Toboso Menudo
HIPPOPHAE	Sea buckthorn
rhamnoides	Sallow thorn
HIRTELLA	
racemosa	Pigeon-plum
HOLODISCUS	
discolor	Ocean spray
HOMALANTHUS	
populifolia	Bleeding-heart tree
HORDEUM	
jubatum var. *caespitosum*	Bobtail barley
HUMULUS	Hops
HYMENAEA	Locust
altissima	Locust
courbaril	West Indian Locust
HYPARRHENIA	
contracta	Thatching grass
HYPHAENE	Gingerbread palm
IDRIA	Boojum
ILEX	Holly
chinensis	Kashi holly
paraguaiensis	Paraguay tea
verticillata	Black alder
IPOMOEA	
batatas	Sweet potato
IRIS	Flag\Fleur-de-Lis
IRVINGIA	Wild mango
JASMINUM	Jasmine
JATROPHA	
curcas	Barbados nut\Physic nut
JUGLANS	Walnut
cathayensis	Chinese butternut
cinerea	Butternut
nigra	Black walnut
regia	English walnut
JULBERNARDIA	
paniculata	Mnondo
JUNIPERUS	Juniper
communis	Common juniper
virginiana	Eastern redcedar
KALMIA	Laurel
angustifolia	Narrow-leaved laurel
latifolia	Mountain laurel
KHAYA	African mahogany
nyasica	Nyasaland mahogany
senegalensis	Red mahogany
KOELREUTERIA	Golden-rain tree
paniculata	Varnish tree

LABURNUM	Bean tree
anagyroides	Golden-chain
LACTUCA	Lettuce
LAGERSTROEMIA	Crapemyrtle
LANTANA	Shrub verbena
camara	Yellow sage
LARIX	Larch
decidua	European larch
laricina	Tamarack
LAURUS	Laurel\Sweet bay
obtusifolia	Patte shoonda
LAWSONIA	
alba	Indian privet
LESPEDEZA	Bush clover
LEUCAENA	
pulverulenta	Giant Leadtree
LEUCOPHYLLUM	
frutescens	Barometer bush\Ceniza
LEUCOSPERMUM	Pincushion flower
LIGUSTRUM	Privet\Trena
japonica	Japanese privet
ovalifolium	California privet
LINDERA	
benzoin	Spicebush
LIQUIDAMBAR	Liquidambar
formosana	Formosan Gum
styraciflua	Sweetgum
LIRIODENDRON	
tulipifera	Tuliptree\White poplar
LITCHI	
chinensis	Litchi
LITSEA	
citrata	Addakurry
polyantha	Soalu
salicifolia	Digloti
LONCHOCARPUS	Cabbagebark
LONICERA	Honeysuckle
japonica	Japanese honeysuckle
LUEHEA	
speciosa	Kazcat
LUPINUS	Lupine
LYCIUM	Wolf-berry\Desert thorn
LYCURUS	
phleoides	Wolftail
LYSILOMA	
auritum	Quiebracha
divaricata	Quiebracha
LYTHRUM	Loosestrife
salicaria	Purple loosestrife
MACHILUS	
bombycina	Som
odoratissima	Laurel\Sum tree
MACLURA	
pomiferum	Osage orange
MAGNOLIA	Magnolia
acuminata	Cucumbertree
grandiflora	Southern magnolia
sphenocarpa	Chapa
tripetala	Umbrella magnolia
virginiana	Sweetbay
MALUS	Apple
coronaria	Sweet crab apple
pumila	Apple
MALVAVISCUS	Sleepy mallow
arboreus	Wax mallow
arboreus var. *drummondii*	Turk's Cap
MANGIFERA	Mango
MANIHOT	Cassava\Tapioca plant
MANILKARA	
chicle	Sapodilla
MAYTENUS	Mayten
MELASTOMA	
malabathricum	Indian rhododendron
MELIA	Bead tree
azedarach	Chinaberry
MELIANTHUS	Honeybush\Persian lilac
MELILOTUS	Sweet clover
alba	White sweet clover
officinalis	Yellow sweet clover
MICHELIA	Champa
figo	Banana shrub
MICONIA	
argentea	Ceniza

MILNEA	
roxburghiana	Punyava
MIMOSA	Mimosa
borealis	Catclaw
dealbata	Silver wattle
lindheimeri	Lindheimer Mimosa
pigra	Zarza
richelmanni	Catclaw
MONOTES	Yellow wood
MORINGA	
oleifera	Horseradish-tree
MORUS	Mulberry
alba	White mulberry
MUEHLENBECKIA	Wire plant
MUHLENBERGIA	Muhly
MUNTINGIA	
calabura	Jamaica-cherry\Strawberry-tree
MYRICA	Bayberry
carolinensis	Bayberry
cerifera	Southern bayberry\Waxmyrtle
conifera	Waxmyrtle
cordifolia	Bayberry
gale	Sweetgale
pensylvanica	Northern bayberry
MYRSINE	
melanophloeos	Rapanea
NEMOPANTHUS	Mountain holly
mucronata	Catberry
NEPHELIUM	
litchi	Lychee
NERIUM	Oleander
oleander	Common oleander
NYSSA	Tupelo
sylvatica	Black gum
OCHNA	Bird's-eye bush
OCIMUM	Basil
OCOTEA	
veraguensis	Pububuk
OLEA	Olive
europaea	Common Olive
paniculata	Olive

OPUNTIA	Cholla\Prickly pear
OSTRYA	Hop hornbeam
virginiana	American hop hornbeam
OXYDENDRUM	
arboreum	Sourwood
PACHYCORMUS	
discolor	Elephant tree
PAEONIA	Peony
PANAX	Ginseng
PAPPEA	
capensis	Wild plum
PARKIA	Locust
biglobosa	African locust
intermedia	Locust bean
speciosa	Locust bean
PARKINSONIA	Jerusalem-thorn
PARTHENOCISSUS	Woodbine
quinquefolia	Virginia creeper
tricuspidata	Boston Ivy
PASSIFLORA	Passionflower
quadrangulare	Grenadine
PAULLINIA	
cururu	Guarana
PELTOPHORUM	
africanum	Wattle
PENNISETUM	Elephant grass
PERSEA	
americana	Avocado
borbonia	Redbay
PHASEOLUS	Bean
vulgaris	Kidney bean
PHELLODENDRON	Cork tree
PHILADELPHUS	Mock orange
PHORADENDRON	Mistletoe
PHOTINIA	
arbutifolia	Christmasberry
PHYLLANTHUS	
emblica	Emblic
PHYSOCARPUS	Ninebark
PICEA	Spruce
glauca	White spruce
marina	Black spruce

PIMPINELLA	Riceflower
saxifraga	Anise
PINUS	Pine
banksiana	Jack pine
caribaea	Slash pine
contorta	Lodgepole pine
contorta var. *latifolia*	Lodgepole pine
echinata	Shortleaf pine
edulis	Pinyon pine
griffithii	Himalayan pine
halepensis	Aleppo pine
jeffreyi	Jeffrey pine
laricio	Black pine
maritima	Corsican pine
massoniana	Red pine
mitis	Shortleaf pine
monophylla	Singleleaf pine
mugo	Mugo pine
nigra	Austrian pine
patula	Mexican yellow pine
ponderosa	Ponderosa\Yellow pine
radiata	Monterey pine
resinosa	Red pine
rigida	Pitch pine
serotina	Pond pine
strobus	Eastern white pine
sylvestris	Scotch pine
taeda	Loblolly pine
thunbergiana	Japanese Black pine
PIPER	Pepper
PISTACIA	Pistachio
chinensis	Pistachio
lentiscus	Mastic tree
terebinthus	Cyprus\Turpentine
PITHECELLOBIUM	Blackbead
flexicaule	Texas-ebony
lanceolatum	Blackbead
saman	Rain tree
PLATANUS	Sycamore
lindeniana	Plane
occidentalis	American plane-tree
orientalis	Oriental sycamore
PLUMERIA	Temple tree
POLYSCIAS	
fulva	Parasol tree
POPULUS	Poplar
alba	White poplar
balsamifera	Balsam poplar
deltoides	Eastern cottonwood
fremontii	Fremont cottonwood
gileadensis	Balm-of-Gilead
grandidentata	Bigtooth aspen
nigra	Black poplar
tremuloides	Quaking aspen
PORLIERIA	
lorentzii	Soap bush\Guayacán
POTENTILLA	Cinquefoil
POTERIUM	Sea holly
PROSOPIS	Mesquite
alba	Mesquite
algarobilla	Mesquite
glandulosa	Honey mesquite
glandulosa var. *torreyana*	Torrey\Western honey mesquite
juliflora	Honey mesquite
PROTEA	Sugar bush
PRUNUS	Cherry\Plum
americana	Domestic plum
amygdalus	Almond tree
armeniaca	Apricot
avium	Bird cherry
cerasus	Sour cherry
demissa	Western chokecherry
domestica	Garden plum
emarginata	Bitter cherry
fasciculata	Desert almond
ilicifolia	Hollyleaf cherry
maritima	Beach cherry
padus	Bird cherry
pensylvanica	Pin cherry
persica	Peach
puddum	Wild black cherry
salicina	Japanese plum
serotina	Wild black cherry
spinosa	Blackthorn\Sloe

virginiana	Chokecherry
PSEUDOTSUGA	Douglas-fir
menziesii	Douglas-fir
PSIDIUM	Guava
guajava	Common guava
PTEROCARPUS	Mukwa
PTEROCARYA	Wing nut
stenoptera	Chinese wing nut
PUNICA	Pomegranate
PURSHIA	
tridentata	Antelope bitter brush
PYRACANTHA	Firethorn
PYRUS	Pear
americana	American mountain ash
communis	Pear
coronaria	American crab apple
malus	Apple
pyrifolia	Oriental pear
QUAMOCLIT	
setosa	Bindweed
QUERCUS	Oak
agrifolia	Coast live oak
alba	White oak
aliena	Oriental white oak
arizonica	Arizona white oak
bicolor	Swamp white oak
cerris	Turkey oak
chihuahuaensis	Chihuahua oak
coccinea	Scarlet oak
dentata	Daimyo oak
dilatata	Green oak of the Himalaya
dumosa	California scrub oak
emoryi	Emory oak
falcata	Southern red oak
fusiformis	Texas live oak
gambelii	Gambel oak
glandulifera	Konara oak
grisea	Gray oak
havardii	Shin\Havard oak
ilex	Holly\Holm oak
ilicifolia	Bear oak
imbricaria	Shingle oak
incana	Grey oak
kelloggii	California black oak
laevis	Turkey oak
lobata	Valley oak
lyrata	Overcup oak
macrocarpa	Bur oak
marilandica	Blackjack oak
michauxii	Swamp chestnut oak
muehlenbergii	Chinkapin oak
myrtifolia	Myrtle oak
nigra	Water oak
nuttalli	Nuttall oak
oblongifolia	Mexican blue oak
palustris	Pin oak
pedunculata	English oak
petraea	Durmast oak
phellos	Willow oak
pini	Chestnut oak
prinoides	Dwarf chestnut oak
prinus	Chestnut oak
robur	English oak
rubra	Northern red oak
serrata	Konara oak
shumardii	Shumard red oak
shumardii var. *texana*	Texas oak
stellata	Post oak
texana	Texas red oak
turbinella	Shrub live oak
undulata	Wavy leaf oak
velutina	Black oak
virginiana	Live oak
QUISQUALIS	
indica	Rangoon creeper
RHAMNUS	Buckthorn
alnifolia	Alder-leaved buckthorn
californicus	California coffee-berry
crocea	Hollyleaf buckthorn
frangula	Alder buckthorn
prunoides	Buckthorn
purshiana	Cascara
RHAPIS	Lady palm
RHIZOPHORA	Mangrove
mangle	American mangrove
RHODODENDRON	Azalea

RHUS	Sumac
angustifolia	Bastard willow
aromatica	Fragrant sumac
cismontana	Mountain sumac
copallina	Dwarf\Winged sumac
crenata	Bastard willow
glabra	Smooth sumac
integrifolia	Mahogany sumac
laevigata	Bastard willow
lancea	Bastard Willow
laurina	Laurel sumac
longifolia	Bastard willow
lucida	Bastard willow
microphylla	Littleleaf sumac
radicans	Poison ivy
tomentosa	Bastard willow
toxicodendron	Poison ivy
trilobata	Skunk-brush
typhina	Staghorn sumac
verniciflua	Wax tree
villosa	Bastard willow
RHYNCHOSIA	
reticulata	Kanacabtsonot
RIBES	Gooseberry\Currant
alpinum	Alpine currant
americanum	Black currant
cereum	Wax currant
cynosbati	Pasture gooseberry
floridum	American black currant
grossularia	English gooseberry
lacustre	Swamp currant
nigrum	Black currant
rubrum	Red currant
sativum	Common currant
RICINUS	
communis	Castor-oil bean
ROBINIA	Locust
neomexicana	New Mexico black locust
pseudoacacia	Black locust\False acacia
ROSA	Rose
multifolia	Rose
RUBUS	Bramble
allegheniensis	Sow-teat blackberry
idaeus	Wild red raspberry
occidentalis	Black raspberry
ulmifolius	Blackberry\Dewberry
RUMEX	Sorrel
SALIX	Willow
alba	European white willow
babylonica	Weeping willow
caprea	Sallow/Goat willow
discolor	Pussy willow
exigua	Sandbar willow
fragilis	Crack willow
hindsiana	Valley willow
humilis	Prairie willow
interior	Sandbar willow
lasiolepis	Arryo willow
longifolia	Sandbar willow
lucida	Shining willow
viminalis	Basket willow
SALVIA	Sage
SAMBUCUS	Elderberry
canadensis	American elderberry
pubens	American red elderberry
racemosa	European red elderberry
SAPINDUS	Soapberry
saponaria	Wingleaf soapberry
SAPIUM	
biloculare	Mexican jumping-bean
sebiferum	Chinese tallow-tree
SARCOSTEMMA	
brevistigma	Soom
SASSAFRAS	
albidum	Sassafras
SCHEFFLERA	Umbrella tree
SCHINUS	Peppertree
molle	California peppertree
terebinthifolius	Brazil peppertree
SCHLEICHERA	
oleosa	Gum-lac\Ceylon oak
SCHOTIA	
brachypetala	Tree Fuchisa
speciosa	Boerboon

SCLERIA
rehmanii Sedge
SCLEROCARYA Marula
caffra Wild mango
SENECIO Groundsel
deltoideus Ragwort
SERENOA Saw palmetto
SESBANIA
grandiflora Agati\Rattlebox
SETARIA
viridis Bristle grass
SHEPHERDIA Buffaloberry
argentea Silver buffaloberry
canadensis Soapberry
SHOREA Sal
SIDEROXYLON
inerme Milkwood
SIMMONDSIA Jojoba
SITANION Squirreltail
jubatum Big squirreltail
SOLANUM Nightshade
hazenii Nightshade
tuberosum Potato
variabile Nightshade
SORBUS Mountain ash
americana American mountain ash
SPARTIUM
junceum Spanish broom
SPATHODEA
campanulata
Flame tree\African tuliptree
SPIRAEA Meadowsweet
alba Meadow spiraea
betulifolia var. *corymbosa*
Dwarf spiraea
latifolia Meadowsweet
menziesii Menzies spiraea
salicifolia Willowleaf spiraea
tomentosa Hardhack
SPIROSTACHYS
africana Sandal wood\Tambootie
SPONDIAS Hogplum
mombin Yellow mombin
purpurea Purple mombin
SPOROBOLUS Cat's tail grass
airoides Alkali sacaton
STYRAX Snowbell
americana American snowbell
SWIETENIA Mahogany
mahogani West Indian mahogany
SYMPHORICARPOS Snowberry
albus Snowberry
mollis Creeping snowberry
oreophilus Mountain snowberry
rotundifolius Roundleaf snowberry
SYMPLOCOS Sweetleaf
crategoides Asiatic sweetleaf
SYRINGA Lilac
vulgaris Common lilac
SYZYGIUM
aqueum Swamp peach
malaccense Malay apple
owariense Water-berry tree
TABEBUIA Trumpet tree
argentea Guayacán
rosea Palo de rosa
TAMARINDUS Tamarind
TAMARIX Tamarisk\Salt cedar
gallica French tamarisk
TAXODIUM Cypress
distichum Bald cypress
TECTONA
grandis Teak
TERMINALIA Assegai wood
bellirica Myrobalan
catappa Indian almond
TETRADYMIA Horsebrush
axillaris Cotton thorn
glabrata Littleleaf horsebrush
spinosa Shortspine horsebrush
TETRANTHERA
diglottica Digluttee
glauca Dighlate
macrophylla Sonhalloo
monopetala Sualu tree
polyantha Mezankuri

TETRASTIGMA	Java grape
TEUCRIUM	Germander
THEA	
sinensis	Tea
THEOBROMA	
cacao	Cocoa
THUJA	Cedar
occidentalis	Northern red cedar
TIBOUCHINA	Glory bush
TILIA	Basswood\Lime\Linden
americana	American basswood
europaea	Common lime
heterophylla	White basswood
platyphyllos	Broad leaved lime
TRICHILIA	Cape mahogany
TRIFOLIUM	Clover
repens	White clover
TSUGA	Hemlock
canadensis	Eastern hemlock
TURNERA	
ulmifolia	Yellow alder
TURPINA	
sphaerocarpa	Pomifera
UAPACA	Wild loquat
ULMUS	Elm
alata	Winged elm
americana	American elm
campestris	English elm
fulva	Slippery elm
procera	English elm
pumila	Dwarf\Siberian elm
racemosa	Rock elm
rubra	Slippery elm
thomasii	Rock elm
URVILLEA	
ulmacea	Apaac
VACCINIUM	Blueberry
myrtillus	Thinleaf huckleberry
VIBURNUM	Arrowwood
cassinoides	Withe rod
dentatum	Southern arrowwood
opulus var. *sterile*	Snowball
trilobum	Highbush cranberry
VERNONIA	Ironweed
VISCUM	Mistletoe
VITIS	Grape
vinifera	European\Wine grape
vulpina	Winter grape
WATSONIA	Bugle lily
WEIGELA	Fairy trumpets
WISTERIA	Wisteria
sinensis	Chinese wisteria
ZANTHOXYLUM	Prickly ash
fagara	Wild lime prickly ash
ZEA	
mays	Cultivated corn\Maize
ZELKOVA	
serrata	Japanese zelkova
ZINNIA	
elegans	Common zinnia
ZIZIPHUS	Jujube
jujuba	Common jujube
mauretanicus	Jujube
mauritiana	Indian jujube
mucronata	Catch thorn\Wait-a-bit thorn
spinachristii	Catch thorn\Wait-a-bit thorn
ZUELANIA	
guidonia	Tamay

APPENDIX II: PERSONAL CORRESPONDENTS

Allen, Michael G., 33 Bridgefield, Farnham, Surrey GU9 8AW, England
Beutler, Larry, 8106 Whitman Drive, Magna, UT 84044 USA
Carr, Thomas W., 6626 Weckerly Drive, Whitehouse, OH 43571 USA
Collins, Michael M., 11901 Miwok Path, Nevada City, CA 95959 USA
Curtis, W. Matthew, 695 Sunrise Boulevard, Prescott, AZ 86301 USA
Evans, Howard E., Colorado State University, Department of Entomology, Fort Collins, CO 80523 USA
Furr, Mecky, 7925 Cross Pike, Germantown, TN 38138 USA
Hansen, Kenneth C., P. O. Box 778, Willow Creek, CA 95573 USA
Hawks, David C., 6466 Cowles Mountain Boulevard, San Diego, CA 92119 USA
Herbin, Daniel, 53, allée des Platanes, F-31320 Pechabou, France
Kondratieff, Boris C., Colorado State University, Department of Entomology, Fort Collins, CO 80523 USA
Lemaire, Claude, La Croix des Baux, F-84220 Gordes, France
Lym, Brian, California Academy of Sciences, Golden Gate Park, San Francisco, CA 94118 USA
Naumann, Stefan, Brandenburgische Str. 10, D-1000 Berlin 31, Germany
Nässig, Wolfgang A., Schumacherstrasse 8, D-6052 Muehlheim 3, Germany
Peigler, Richard S., Zoology Department, Denver Museum of Natural History, 2001 Colorado Boulevard, Denver, CO 80205-5798 USA
Probst, Helmut, Augsburger Str. 7, D-8939 Türkheim, Germany
Purdum, David A., 2422 Montvale Road, Roanoke, VA 24015 USA
Rawlins, John E. The Carnegie Museum of Natural History, Department of Invertebrate Zoology, 4400 Forbes Avenue, Pittsburgh, PA 15213 USA
Richards, John L., 10364 South 360 East, Sandy, UT 84070 USA
Ringel, Gerhard, 2256 West Cliff Drive, Santa Cruz, CA 95060 USA
Savage, Patrick J., P. O. Box 1412, St. George, UT 84770 USA
Simon, Leroy D., 2215 Hialeah, Leesburg, FL 32748 USA
Smith, Michael J., 7428 Holworthy Way, Sacramento, CA 95842 USA
Sblendorio, Daniel C., 13 Highview Circle, Middletown, NJ 07748 USA
Tuskes, Paul M., 3808 Sioux Avenue, San Diego, CA 92117 USA
Tuttle, James P., 3838 Fernleigh Avenue, Troy, MI 48083 USA
Vuattoux, Robert, 409, Chemin des Caillades, F-06480 La Colle sur Loup, France
Warren, Andrew D., 9951 East Ida Place, Englewood, CO 80011 USA
Williams, Benjamin D. III, P. O. Box 211, Day Road, Pomfret Center, CT 06259 USA
Wilson, Michael F., Cochise County Health Services, P. O. Drawer 1858, Bisbee, AZ 85603 USA
Wolfe, Kirby L., 3090 Cordrey Drive, Escondido, CA 92025 USA

INDEX TO SATURNIID GENERA AND SPECIES

Principle entries are listed in **boldface**.
The use of a hyphen "-" between two page numbers following a species indicates reference to that species is found on all those pages.

INDEX TO SATURNIID FOODPLANTS

Principle entries are listed in **boldface**.
The use of a hyphen "-" between two page numbers following a species indicates reference to that species is found on all those pages.